Program Your Calculator

Program Your Calculator

Gerald R. Rising

Eileen K. Schoaff

Deborah Moore-Russo

William R. Parks
Stanwrite@aol.com

www.wrparks.com

Consultants

Mark Spahn

David Patrick

Julie Sarama

Douglas Clements

Steve West

The cover is a modified photo of the bronze sculpture of "The Thinker" by Auguste Rodin, whose first cast, of 1902, is now in the Rodin Museum in Paris.

Contents

Chapter 1.
Introduction

*Most of you are familiar
with the virtues of a programmer.
There are three, of course:
laziness, impatience, and hubris.*
 – Larry Wall

When computers and programmable calculators first became available to schools, it was accepted that students would understand mathematical concepts better if they learned how to translate those ideas into computer or calculator programs. Sadly, in one of those pendulum swings that beset education, that idea has receded from educational thinking, and teaching even the rudiments of programming has disappeared from school math programs. Thus most college students come equipped with no programming skills. If your experience differs from this, consider yourself very fortunate.

You may, however, have used a calculator in your school math program. If you did, that use was almost certainly limited to drawing graphs and carrying out calculations that could be done on much cheaper scientific calculators. This is, I suggest, a sad state of affairs.

Calculators like the TI-84 for which this text is designed[1] accept programs in a modified form of BASIC, the simplest programming language.[2] Simple, yes, but BASIC is a powerful language. It was developed by senior mathematician John

Kemeny and his colleague Thomas Kurtz. Kemeny was chair of the Dartmouth College mathematics department until he became president of that institution. (He was later chair of the commission that identified the reasons for the *Challenger* accident that killed six astronauts.)

It is important to understand that those calculators differ even from other Texas Instruments calculators like the TI-89, TI-92 and TI-Nspire, but the various models in the TI-84 series differ almost entirely in memory. For users interested in developing large programs, the more memory the better, even though, as you will learn, to save memory, programs can be archived or transferred to a computer for storage and retrieval.[3]

Too many calculators play the same kind of role for their owners as do coffee table books: they are admired but rarely accessed. In this text the calculator will be used regularly to solve problems, to illustrate concepts, and, most important to you, to short-cut computation. You will of necessity become familiar with many of your calculator's special features, but even those who use calculators constantly continue to find both additional features and minor glitches that require work-arounds. The following introduction will not then make you an expert; only use will make you proficient working with your instrument. What you need to do if you get stuck is refer back to this chapter, to your calculator manual, or to your instructor for guidance.

All of the following instructions assume just two things: that you have turned on your calculator and that you know how to press appropriate keys to carry out operations — for example, **3.2 × 5.7 ENTER** to multiply 3.2 by 5.7 (and get 18.24); **5.39 x^2 ENTER** to square 5.39 (29.0521); **log 5) ENTER** to display the common logarithm of 5 (.6989700043); and **2ND** √ **5)** **ENTER** to find the square root of 5 to ten digit accuracy (2.236067977). I urge you to stop here briefly to check those computations.

---

1 There are a number of programmable graphing calculators, each with its own idiosyncrasies. With minor modifications, later models of the TI-83 series may also be used for this text. The most significant difference relates to the timing feature that was added to the TI-84. Any program steps involving timing must be removed from TI-84 programs to make them work on the TI-83.

2 Some would argue that the programming language Logo is still simpler and you may even have met this language in elementary school. For some applications I would agree, but here the case is moot. Your calculator does not program with Logo.

3 A partial alternative to this chapter for the Casio FX series of calculators may be downloaded from the web at: `www.buffalo.edu/~insrisg/LINKS/CasioInt.htm`. I invite Casio users to submit translations of the programs of this text for posting there.

Chapter 2.
Getting Started

Computers are good at following instructions
but not at reading your mind.
—Donald Knuth

Before you begin writing programs, you should familiarize yourself with a number of the keys you may not have employed before, but that are of general use in working with your calculator. The exercises will give you practice with them.

ENTER is like the **RETURN** key on your computer. It means: Carry out the preceding operation or, in a program: Include this instruction and make room for another. This will be the key you will use most often. On some inexpensive calculators it is replaced by an = key, but the relation = plays a very different role in your calculator.

CLEAR erases the screen or in a program erases the current program line.

DEL erases the current character or instruction. The current character is the one on which the cursor, that black rectangle, is blinking. It also erases a blank program line.

STO> is arguably the most important key on your calculator. In programs it appears as an arrow, →. It allows you to store a value in a location named after the indicated key. For example, **5 STO> ALPHA T ENTER** will store **5** in location **T** and **5** remains the value of **T** until some other number is stored there. (This is something you should be aware of as that **5** will remain stored in **T** even if your calculator is turned off and then on again.) Now if you press **ALPHA T ENTER**, that stored value, 5, will appear on your screen. You can also use this key to store the results of calculations. For example, when your calculator is in Degree mode, `sin(30) STO> ALPHA V` will store the value, **.5**, in **V**.

(-) This is a troublesome key. You must be careful to distinguish between subtraction which uses the minus key, **-**, and negative numbers which use the **(-)** key. Thus you would write **5-3**, but **(-)3+5**.

The scrolling keys. Near the upper right of your keyboard is a group of keys marked with triangles pointing left, right, up and down. These keys are useful in moving the cursor from line to line and back and forth within lines on your screen.

2ND followed by a key gives you the left-hand instruction above the key. Three examples: to access the instruction INS (to insert a programming keystroke), press **2ND** then **DEL**; to write the value of π to your screen, press **2ND** then π, and to turn your calculator off, press **2ND** then **ON**. (Always press keys in sequence; NEVER press two keys at the same time.)

ALPHA is like **2ND** but it gives you the right-hand instruction above the key. These are mostly letters, as you would expect given the name of the key. If you want to type an **A**, for example, press **ALPHA** then **MATH**. This key only works for the current key then reverts back to normal operation. To shift to this form for several keys in succession, press **2ND** then ALPHA (for **A-LOCK**) ; then to shift back, press **ALPHA** again.

Notation: From now on I will write a key value that is

obtained by **2ND** or **ALPHA** without indicating those keys. Thus when, for example, I write **TEST**, you must realize to access that key you must press **2ND MATH**.

MODE gives you access to the way your calculator is formatted. I suggest that you leave all but one key in the leftmost column darkened. The exception is **Degree**. To make a change, scroll to the desired format and press **ENTER**. When you are finished with this screen press **QUIT**.

INS is a very useful key. In normal entry mode, you type over an entry. For example, if you move your cursor to a **3** and type **2**, the **2** will replace the **3**. To change to "Insert" mode, press the **INS** key. Now what you type will be inserted. For example, suppose again our cursor is on a **3** and we press **INS 27**. The **3** would be moved to the right and the **27** appear in front of it as **273**. When programming, to add a space for a program line, use **INS ENTER** at the beginning of the line you wish to create. You exit this mode by keying **INS** again or by moving the cursor.

QUIT is also useful as an exit or escape. Unfortunately it will take you out of a program you are editing. If you use it then, you must re-access the program to continue. This instruction does not erase anything. (Sometimes you can return to the previous programming screen by keying **CLEAR**.)

That may seem like a great many keys to learn. You'll find, however, that they will soon become familiar through use.

2.1 Exercises

(2.1) Store the value 5 in **A**, 4 in **B** and 54700816 in **C**. Then use these values to calculate the following exercises. Be sure your answers make sense:

(a) **A÷B**   (b) **B/A**   (c) $\sqrt{C}$

(d) $\mathbf{A^B}$ (e) $\mathbf{B^{-A}}$ (f) **Aπ**

(g) `log B` (h) $\mathbf{e^A}$ (i) `log 10`$^\mathbf{B}$

(2.2) Starting with the values stored in exercise (2.1), store **A** in **C**, **B** in **D**, and then **D** in **A**. Once this is done, first guess and then check the values of **A**, **B**, **C** and **D**. Note: All you need to do to find the contents of a storage location is type that letter on the screen and type **ENTER**.

(2.3) Which key would you use to indicate the minus signs for:

(a) **A-B**   (b) **-B**   (c) **-237-239**

(2.4) Mathematician David Feinberg wrote the following brief poem which he titled appropriately "The Square Root of Three":

> *Exactly one-half of*
> *2π-e*
> *Is about three percent more*
> *Than the square root of three.*

Check Feinberg's arithmetic by:

(a) Storing (2π-e)/2 in **X**,

(b) Storing $\sqrt{3}$ in **Y**

(c) Dividing **X** by **Y**.

(d) Your answer in (c) will be a number of the form 1 + a decimal. Convert that decimal to a percent to be compared with 3.

(2.5) A seldom-used blue key is labeled **EE** (to distinguish it from the green key for the letter **E**. The **EE** key is used for scientific notation, but it can also short-cut other representations. Find what this key represents by entering the following:

(a) **EE 1**   (b) **8.73 EE 5**

(c) **2.3 EE -3**

Chapter 3.
A First Program

The [programmer's] main challenge is
not to get confused by the complexities
of his own making.
—Edsger Dijkstra

I live near the Canadian border and in Canada (as well as the rest of the world[1]) they use the metric system. When I travel to Canada and other foreign countries, I find the temperature given in Celsius rather than in Fahrenheit as it is in the United States. You may remember from a school science class that there are simple formulas that allow you to convert from one of these systems to the other. Here is the one that converts from Celsius to Fahrenheit:

$$F = \frac{9}{5}C + 32$$

If you hear a weather report that tells you the Celsius temperature is 20°, you can plug that 20 into the formula, do the calculation and produce F = 68, so the temperature is 68° in our commonly used scale. Please stop here and use your calculator to check that computation.

You probably keyed something like: **9 ÷ 5 ← 20 + 32 ENTER.**[2]

Now suppose you have a great many Fahrenheit values to convert to Celsius. You can, of course, simply plug each one into the formula and do the calculation, but we can develop a program that will simplify that task. Doing so the first time will seem like much work to accomplish little, but you will be left with a tool that will not only do the computation but will remember the formula.

So let's get going. Carefully follow these instructions:

1. Press **PRGM**, scroll to **NEW**, and press **ENTER**.

2. Now you are asked for a name of your program. Let's call it **CTOF**, abbreviating Celsius to Fahrenheit. To enter this name, simply type **C T O F** and then **ENTER**. Note that you did NOT have to change to **ALPHA** mode to do this.

3. You are now ready to enter your program steps. Your first task is to get a value of C into the program. To do so, press **PRGM**, scroll right to **I/O** (for input/output) and down to **2:Prompt**. Then press **ENTER**. (You could have shortened those last two steps by simply pressing **2**.)

4. Type **C**. (This time you do need to use the **ALPHA** key.) When you have the program line, **:Prompt C**, press **ENTER**.

5. At this point, when you run this program, it will ask for a value of C. We now want to convert that value to F using the formula. So type

   9 ÷ 5 ← C + 32 STO> F. And when you have the program line **:9/5*C+32→F**, press **ENTER**.

6. Finally, you wish to report your calculated value of F in the last program line. To do so, press **PRGM**, scroll right to **I/O** again, scroll down to **3:Disp** and press **ENTER**. Key **F** to complete the program line, **:Disp F** and press **ENTER**.

You have now entered the three-line program that should appear like this:

```
PROGRAM:CTOF
:Prompt C
:9/5*C+32→F
:Disp F
```

Press **QUIT** and you will have stored this program in your calculator.

And that's it. Yes, it was a lot of work to develop a three line program, but you have to start somewhere and those seemingly complicated steps will become routine as you enter additional programs. You will see the roles of those program lines when you see the program being run.

Let's see how to do just that. You will be happy to find that running a program is far easier than entering it in your calculator. To use your program **CTOF**, press **PRGM**, then scroll (if necessary) to **CTOF**. Press **ENTER** to call up the program and **ENTER** again to run it. When you do this, your screen should show **C=?** and you can type in any Celsius value. Try 100, the boiling point of water in Celsius. Type in **100 ENTER** and your screen should appear like this:

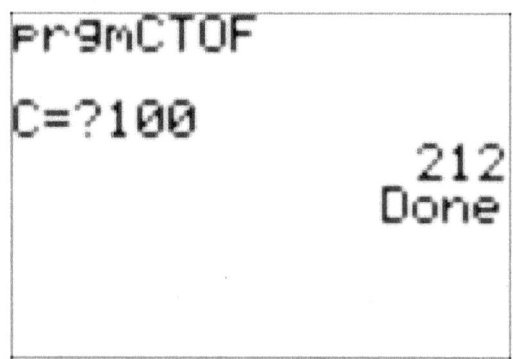

Figure 3.1 CTOF output

Easy enough. And you don't have to go through those steps again if you are careful. Press **ENTER** now and your calculator will ask for another C value. Try 0, the freezing point for water on the Celsius scale.

I'll summarize in highly abbreviated form for your reference, if you should need it, what we did to enter a program and to run a program.

To enter a program:

1. **PRGM NEW**, name your program, **ENTER**.

2. List your program lines, **ENTER** after each one.

3. **QUIT**

To run a program:

1. **PRGM**, scroll to your program name, **ENTER ENTER**.

2. If asked to enter values, do so, pressing **ENTER** after each one.

Entering and running a program should seem easier when the steps are listed that way.

- - - -

1 Well, not quite. If you travel to Myanmar (Burma) or Liberia you may not feel at home but you will find our measurement system in use.

2 The order of operations comes into play here. Your calculator carries out computations by so-called algebraic processing. It does powers and roots, then multiplication and division, then addition and subtraction, all left to right. Here multiplication and division are being processed before the addition and within those processes from left to right.

3.1 Exercises

(3.1) Use your program **CTOF** to convert the following Celsius temperatures to Fahrenheit:

   (a) 37° C: normal body temperature

(b) -273.15° C: absolute zero

(c) 280° C: gasoline will self-ignite

(d) 6000° C: surface temperature of the sun

(e) 107° C: daytime temperature on the moon

(f) -153° C: nighttime temperature on the moon

(3.2) By experimenting (if you don't already know), find the temperature at which Celsius and Fahrenheit are the same. Hint: The temperature is negative.

(3.3) The formula for changing Fahrenheit to Celsius is:

$$C = \frac{5}{9}(F - 32)$$

(a) Develop a program **FTOC** that uses this formula to make this change. Use the format of your **CTOF** program to do so.

(b) Use your program to check your answers in (3.1)(a) and (b) by reconverting those Fahrenheit results back to Celsius.

(3.4) Develop and run a one-line program, **NAME**, that prints your name on your calculator screen. (Hint: Modify the third line of your CTOF program, placing the letters of your name in quotes.)

(3.5) Develop a three-line program **TRIGSUM**, modeled on the **CTOF** program, that calculates values for the function y = sin x + cos x. Use your program to evaluate:

(a) sin 23° + cos 23°

(b) sin 245° + cos 245°

(c) sin 720° + cos 720°

Chapter 4.
Editing Programs

Debugging is twice as hard
as writing code in the first place.
—Brian Kernighan and P.J.Plauger

Program editing is a very important aspect of programming. There are many reasons for this. Here are two:

1. It is extremely easy to make mistakes when entering program steps. You must remember that your calculator is a machine. It doesn't think. To your calculator, "nearly correct" is simply "wrong" and your calculator will produce an error message or, even worse, an incorrect and misleading answer. When your program does not work, you must locate your error (not always an easy task) and correct it. Searching out and correcting program mistakes is called debugging.[1]

2. You may have a perfectly good program that you want to do more for you.

To edit a program that you have already entered, key **PRGM**, scroll right to **EDIT**, down to your program name, and then **ENTER**. To illustrate the editing process, we'll edit the program **CTOF** that we developed in the last section.

```
PROGRAM:CTOF
:Prompt C
:9/5*C+32→F
:Disp F
```

Before continuing, a word about the role of the colon (:) in your program is in order here.[2] You didn't have to type them, because each time you pressed **ENTER** your program inserted one on a new line. Those colons separate instructions and you can combine instructions in the same line by using them. For example, you could have written your program CTOF all on one line as:

PROGRAM:CTOF
:Prompt C:9/5*C+32→F:Disp F

I urge you not to do that, because the line-by-line form is easier to follow and edit and you are not saving any calculator memory by doing so. Internally, your calculator processes your program in this short form: to it those new lines don't matter. From now on, I am going to omit those colons when displaying programs.

Make some minor changes in the program **CTOF**. First, modify that first line **Prompt C** to make it convey more information to the user. In editing mode scroll to that first line and press **CLEAR.** As you might expect, that erases that line, leaving it blank. Replace that line with the instruction:

Input "CELSIUS TEMP? ",C

Typing that will seem more complicated than it is. Find **Input** at **PRGM I/O**. Then use **A-LOCK** to type the characters up to the second ". (The space character is on the **0** key, the " is on the **+** key and the **?** is on the **(-)** key.) Then type **ALPHA** to unlock the **ALPHA** format, add the comma and finally the letter **C**.

Quit and run your program **CTOF**. At the outset, instead of displaying **C=?**, it will display **CELSIUS TEMP?** . This way it informs the user what is to be input. While the difference between these two input instructions here is not too great, in other programs it may be significant.

You now know how to change program input. Next we'll make changes in program output. Go into **EDIT** mode and scroll to the output instruction line, **Disp F**. Recall that this instruction simply displayed a number on your screen, like **32**, and showed **Done** on the line after that. We'll add some lines that will further identify that output.

Here is the program you can create:

```
PROGRAM:CTOF
Input "CELSIUS TEMP? ",C
9/5*C+32→F
Disp "FAHRENHEIT TEMP"
Disp F
Disp "DEGREES"
```

To make these changes, you will want to enter new lines. To enter the line before the **Disp F** that is already in your program, move your cursor to the beginning of that line and key **INS ENTER**. Bingo: the blank line appears. Then move your cursor up to enter the new line.

Now if you run that program with an input of **30**, it will display the following:

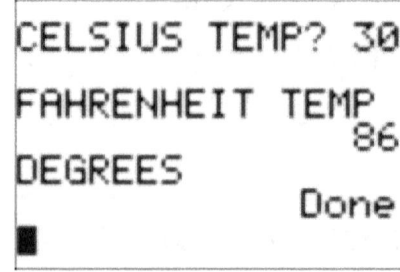

Figure 4.1 CTOF output when modified

That may convey the information you want to display but it is very ugly and you can do much better using the **Output** command. It introduces a few complications, however. Once you understand the coordinates of that home screen you will be able to locate information on it.

Your home screen is assigned coordinates in the form (row,column) with the rows running from 1 to 8 and the columns from 1 to 16.

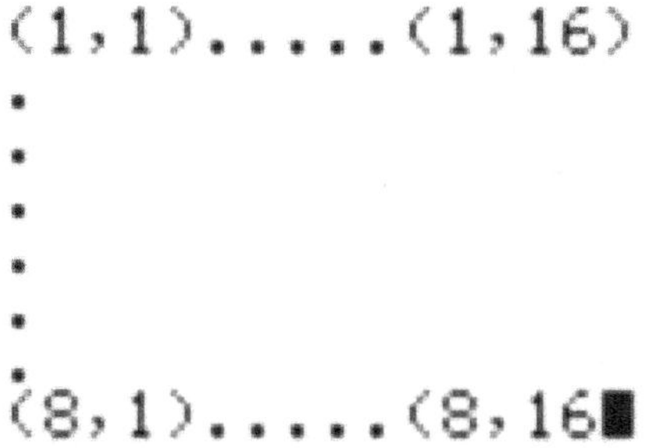

```
(1,1).....(1,16)
  .
  .
  .
  .
  .
  .
(8,1).....(8,16█
```

Figure 4.2 Home Screen Coordinate Range

With the **Output** command, you can locate information on your home screen using these coordinates. For example, the command **Output(1,1,"HELLO")** would locate the word **HELLO** in the upper left corner of your screen and the command **Output(8,10,N)** would place the current value of **N** in the lower right part of your screen. A little arithmetic, 8 ← 16, will show that you have 128 locations on this screen.

If you recall, we modified the program CTOF to provide more input and output information, but the output information was oddly placed on the screen. I am going to revise that program to replace the lines

```
Disp "FAHRENHEIT TEMP"
Disp F
Disp "DEGREES"
```

with output lines that provide better format. Here is the resulting program:

```
PROGRAM:CTOF
Input "CELSIUS TEMP? ",C
9/5*C+32→F
ClrHome
Output (3,1," CELSIUS TEMP")
Output (4,4,C)
Output (4,7,"DEGREES ")
Output (6,1,"FAHRENHEIT TEMP ")
Output (7,4,F)
Output (7,7,"DEGREES")
```

Run that program and you will produce a much neater output. Your screen should look like this:

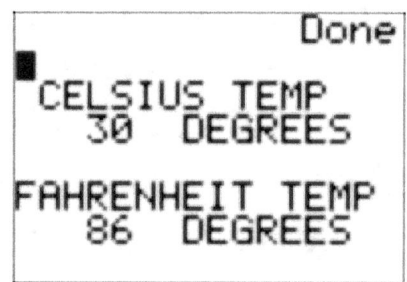

Figure 4.3 Output of Latest Version of CTOF

---

1 The word debugging arose from an experience by early programmer Grace Hopper who found a real bug, a moth, in her computer that was causing her problems. Dr. Hopper went on to become a U. S. Navy admiral.

2 The **:** is the **ALPHA** value of the decimal point key.

4.1 Exercises

(4.1) Edit the program **NAME** you created in exercise (2.4) so that it prints **OWNER** on the line after your name.

(4.2) Edit the program **FTOC** you created in exercise (2.3) so that it identifies input and output.

*

Comment: In the following chapters I will be introducing tools that give you quite remarkable power over your calculator. In doing so I will use simple or even trivial examples. These exercises are chosen to familiarize you with the structures with as few complications as possible. The final chapter will then use those tools to address many more interesting and involved situations.

Chapter 5.

Control Structure 1:

`Goto...Lbl`

> *Go to heaven for the climate,*
> *hell for the company.*
> – Mark Twain

Remember how you learned the order of operations in school algebra courses. In computing the value of an expression you follow the order: (1) powers and roots, (2) multiplication and division, and finally (3) addition and subtraction. You can adjust that order by using parentheses. For example, $6 \times 2 + 5 = 12 + 5 = 17$, but $6 \times (2 + 5) = 6 \times 7 = 42$, because the parentheses require that you process that operation first. (All calculators today follow this so-called algebraic order.[1])

Your calculator also has rules for processing the program steps you have entered. The basic rule is: Process the steps in the order they appear. Thus it processes the first line, then the second, then the third, and so on until it has completed the program lines, at which point it stops and records **Done.**

Control structures, like those parentheses in algebraic simplification, allow you to modify the order in which program steps are processed. As you will see later, they also allow your program to make decisions.

Edit your **CTOF** program by adding **Lbl A** at the beginning and **Goto A** at the end. You will find **Lbl** and **Goto** by scrolling down in **PRGM**.

```
PROGRAM:CTOF
Lbl A
  Input "CELSIUS TEMP? ",C
  9/5*C+32 →F
  Disp "FAHRENHEIT TEMP"
  Disp F
  Disp "DEGREES"
Goto A
```

Here for the first time I am using indentation to clarify the structure of your program. In this case the program lines 3-7 are indented because they are controlled by the **Lbl** and **Goto** instructions. Indentation like this is standard computer science practice and I will continue to do this. It is important to understand, however, that that indentation does NOT occur in your program. Your calculator not only does not recognize indentation, but it will not accept it.

When your calculator processes that program, it first encounters that **Lbl A** line and records internally the location of that line. Then it processes the following lines as it did before. When it comes to the last line, however, that **Goto A** transfers processing back to that **Lbl A** and the program repeats by asking for a new value for **C**. Run this program to see how this works.

Although this did not accomplish much more than the original program, it shows you how **Goto** changes the processing order.[2] In fact, its use here also creates a problem. You have created what is called an infinite loop. There is no exit from this loop, no opportunity for the calculator to finish and report **Done**. To "break" an infinite loop, key **ON**. It will give you an opportunity to quit. You'll get an error message, but at least you broke out of that loop. Such loops do not represent good programming and you will learn later how to avoid them.

---

1 This was not true when calculators were first being produced in the 1970s. Some simply calculated from left to right and still others followed so-called reverse Polish or Lucasiewicz order from right to left.

2 Early programmers overused this control structure until the Dutch computer scientist Edsger Dijkstra in 1968 called for its abolition. He claimed that the use of many **Goto** instructions complicated the task of analyzing and verifying programs. The **Goto** command remains useful, however, and I introduce it first here because of its simplicity.

5.1 Exercises

(5.1) **Goto** and **Lbl** instructions can appear anywhere in your program.

  (a) To show this, add the instruction **Disp "CONVERT TEMPS"** to the beginning of your **CTOF** program.

  (b) Run the resulting program several times. Why didn't the **CONVERT TEMPS** output appear after the first run?

(5.2) Edit your program **NAME** by adding **Lbl 1** to the beginning and **Goto 1** to the end. Run this program. What is happening?

Chapter 6.

Control Structure 2:

`If...Then...Else...End`

*Programmers are in a race with the Universe
to create bigger and better idiot-proof programs,
while the Universe is trying
to create bigger and better idiots.
So far the Universe is winning.*
– Rich Cook

Your first control structure merely changed your route through the program. This next one forces your program to make decisions based on prior information.

Consider the first of these structures by means of a simple example. Suppose we want our calculator to solve equations of the form

$$y = \frac{2x+4}{10-5x}$$

for various values of x.

That is a straightforward exercise. We could simply enter the program:

```
PROGRAM:EVALUATE
Prompt X
(2X+4)/(10-5X) →Y
Disp Y
```

There is a small problem with this program, however. Notice

what would happen if you enter **2** for **x**. The denominator of the fraction becomes **0** and the function is undefined. If you ran the program and entered **2** for **X** at the prompt, you would get an error message and the program would stop.

One way to avoid this is to modify the program in the following way (the **=** symbol is found using **TEST**):

```
PROGRAM:EQUATION
Prompt X
If X=2
Then
   Disp "CALCULATION UNDEFINED"
Else
   (2X+4)/(10-5X)→Y
   Disp Y
End
```

That program may seem complicated to you, but it is very close to our natural way of saying things. You might say, for example, "If you loan me $5, then I will be your friend, or else I won't like you." If you omit the "or" and write that in the form of a pseudo-program you would have:

```
If you loan me $5
then
   I will be your friend
else
   I won't like you.
```

The End in the program informs the calculator that you have completed the structure.

Here is how this works in general:

```
If true-false test question
Then
   (when true) DO THIS
Else
   (when false) DO THIS
End
```

Of course, your calculator won't be responding to tests like whether or not you are loaned $5. Instead, the kind of test questions to which it responds are mathematical, like **A>B** or, as in our example, **X=2**. Be sure to notice that when you write **X=2** it is a test question. You must get used to the fact that your calculator always considers the **=** sign, like the other signs, **>**, **<**, **≥**, **≤** and **≠** as part of a test.

There are shorter forms of this **If...Then...Else...End** control structure. The following simple examples illustrate these forms.

Here is a program that uses both **Then** and **Else**:

```
PROGRAM:GRADE
Input "TEST SCORE? ",T
If T≥90
Then
  Disp "HONOR GRADE"
  Disp "GOOD WORK"
Else
  Disp "PLEASE TRY HARDER"
End
```

You might choose, however, not to say anything about grades

below 90. Then you could omit the **Else** section of the program.

Stop! Before you enter the following very similar program, I will show you a trick that will save you much retyping. You can call up the program steps of **GRADE** into a new program **GRADE2** in the following way:

1. Begin your new program with **PRGM New** and name it **GRADE2**.

2. At the first program line key **RCL**, then **PRGM**, scroll to **EXEC** and choose the name of your earlier program, **GRADE**.

3. Key **ENTER** twice and the program lines of **GRADE** will appear in **GRADE2** ready for editing.

```
PROGRAM:GRADE2
Input "TEST SCORE ?",T
If T≥90
Then
   Disp "HONOR GRADE"
   Disp "GOOD WORK"
End
```

And finally, if there is just one command to carry out when the test is true, both the **Then** and the **End** can be omitted. In that case you could have the still shorter program:

```
PROGRAM:GRADE3
Input "TEST SCORE? ",T
If T≥90
   Disp "GOOD WORK"
```

6.1 Exercises

(6.1) What is the difference between **X=Y** and **X→Y**?

(6.2) Your calculator doesn't store the words "true" and "false" in answer to tests. See what it does store by entering the following tests on your home screen:

(a) **5=3**   (b) **5>3**

(c) **12*5=30*2**   (d) **4≤1+2.5**

(e) **35-41≥0**   (f) **sin(30)=0.5**

(6.3) Would there be any difference in the output of these two programs?

```
PROGRAM:WATCH1
Input "FTEMP? ",F
If F>102
  Disp "CALL NURSE"
```

```
PROGRAM:WATCH2
Input "FTEMP? ",F
If F>102
Then
  Disp "CALL NURSE"
End
```

(6.4) Here is a program that will round measures to the nearest integer. Copy it into your calculator. Find **iPart(** and **fPart(** under **MATH NUM**.

```
PROGRAM:ROUNDINT
Input "MEASURE? ",M
iPart(M)→N
fPart(M)→D
If D≥.5
```

```
    N+1→N
Disp N
```

(a) Find **iPart(23.35)** and **fPart(23.35)** and describe what those two functions do.

(b) Use the program **ROUNDINT** for the following measures:
      (1) 3.45   (2) 4.91   (3) 5.44
      (4) 3.45 + 4.91 + 5.44

(c) Why don't your answers to (1)-(3) add up to the answer to (4)?

(d) Does the function **iPart(X+.5)** always give the same value as when you process the value **X** in **ROUNDINT**? Why or why not?

(6.5) Create the program **GRADE4** by calling up the program **GRADE** using the **RCL** trick, and then editing it. *Note that the 13 lines are numbered only for reference.*

```
1   PROGRAM:GRADE4
2   Input "TEST SCORE ?",T
3   If T≥90
4   Then
5     Disp "HONORS"
6   Else
7     If T≥60
8     Then
9       Disp "PASS"
10    Else
11      Disp "FAIL"
12    End
```

`13` **End**

This is a case when one **If...Then...Else...End** is embedded in another. In this example, lines `7` through `12` are embedded in the **Else** clause of the basic **If...Then...Else...End**.

We can follow how an individual score would be processed. For example, suppose we input the score **78**. At line `3` the test would be false, because $78 < 90$. That means we skip the **Then** steps `4` and `5` and enter the **Else** at step `6`. In step `7`, since $78 > 60$, the answer is true. That means that we perform the **Then** lines `8` and `9`. In line `9` the program would display **PASS**. We would then skip the **Else** lines `11` and `12`. The program would be complete. Thus it would print **PASS** and **Done** on successive lines.

(a) Describe what happens with an input of **91**.

(b) Describe what happens with an input of **56**.

(c) Compare your answers with what you get running the program **GRADE4**.

Chapter 7.

Some Useful

Mathematical Functions

As soon as we started programming,
we found to our surprise
that it wasn't as easy to get programs right
as we had thought.
– Maurice Wilkes

In addition to the mathematical operations and functions that appear on your calculator keys — x^2, x^{-1}, $\mid$, 10^x, e^x, **LOG**, **LN** and the trig functions — there are a number of useful function keys found by keying **MATH NUM**. Many of these keys represent short-cuts: that is, you could accomplish the same thing in several steps or by simply using other keys, but they do save time in calculating and especially steps in programming. It will be useful for you to become thoroughly familiar with these keys for these reasons.

abs (In mathematics notation **abs(x)** appears as $\mid x \mid$. In elementary school terms, **abs(x)** is "**x** without the sign." Thus **abs(2)** produces **2** and **abs(-2)** `also` `produces` **2**. While you can think of the function in that way, it is more appropriately stated as:

$$|x| = \begin{cases} x, \text{ if } x \geq 0 \\ -x, \text{ if } x < 0 \end{cases}$$

Notice that the program line **abs(x)** replaces these two lines:

If x < 0

$$^-x \rightarrow x$$

int(This is the "greatest integer less than" function in mathematics and **int(x)** is displayed in mathematics as [x] or increasingly today, following computer science usage, as $\lfloor x \rfloor$. In colloquial terms, this function "rounds down" and for that reason is often called the "floor" function. Here are some examples: entering **int(5.9)** will produce **5** and **int(-5.9)** will produce **−6**. This may seem like a very simple tool, but it turns out to be both powerful and useful. As you will see in the exercises, the keys **iPart(, fPart(, round(,** and **remainder(** can all be defined in terms of **int(.**

iPart(This function gives you the integer part of a number. In effect, it strips off any values after the decimal point. Isn't this what **int(** does? Not quite. Compare the following values with those for **int(: iPart(5.9)** will produce **5** just as **int(5.9)** did but **iPart(-5.9)** will produce **−5**, whereas **int(-5.9)** produced **−6**.

fPart(Turn about is fair play. Just as **iPart(** strips off the decimals, **fPart(** strips off the integers. Thus, **fPart(123.456789)** is **.456789**.

round(This function rounds a value to a stated number of decimal places, 0 to 9. For example, **round(123.456789,2)** would produce **123.46**. Notice that the usual rules for rounding apply here: if the next decimal place (the 3rd in this case) is 5 or more, you round up, otherwise round down. In this example, the next decimal place after the 2nd is 6, therefore the 2nd decimal place is rounded up from 5 to 6.

remainder(Think back to elementary school when you first learned division. You divided 17 by 3 to get a quotient of 5 and a remainder of 2. (Later you learned to use that 2 in a fraction to give the quotient 5 2/3.) In programming it turns out that that remainder is useful. For this same example, you would enter **remainder(17,3)** to get **2**. Thus, **remainder(** is answering the question: what is the remainder when the first number is divided by the second number.

There are additional useful functions found in **MATH NUM**, but these are the ones that will turn out to serve you best in programming.

Randomizing Functions

Many mathematical activities require the use of random numbers. Random events are a central concern of probability, but they turn out to be useful in many other applications as well. Your calculator has a random number generator that is very useful in programming. The two forms that are of use to us are both found with **MATH PRB**.

rand Enter this key and you produce a number between 0 and 1. For example, I just keyed **rand ENTER** and the calculator displayed **.2209784733**. Pressing **ENTER** again several times gave **.3694814382**, **.0078387869**, and **.9351587791**. Getting a bunch of random ten-digit numbers like that may not seem very useful to you, but, exactly because they are random, they do indeed serve you well.

randInt(This function produces an integer within a range you specify. Suppose you wish to produce integers in the range $-2 \leq x \leq 5$. You would key **randInt(-2,5)**. I just did that and a series of **ENTER**s produced **-1 0 5 -2 3 3 4 1**.

Some uses of these functions will be suggested in exercises (7.6) to (7.8).

7.1 Exercises

(7.1) For each of the following, determine the answer you believe the calculator function will produce and then check it with that function:

(a) **abs(3.76)** (b) **abs(-3.76)**

(c) **int(π)**　(d) **int(-π)**

(e) **iPart(-π)**

(f) **remainder(37,7)**

(g) **37.65-int(37.65)**

(h) **int(37.65)-37.65**

(i) **round(fPart(π),3)**

(7.2) Enter the following program **TEST** and use it to demonstrate that the two instructions, **fPart(A)** and **A-iPart(A)** are equivalent (that is, they produce the same answers for the same input) by running it for the input values of A listed in (a)-(d).

PROGRAM:TEST

```
Prompt A
Disp fPart(A)
Disp A-iPart(A)
```

(a) **23.7** (b) **-16.46** (c) **-2.1**

(d) **17.8** (e) **.3535**

(7.3) Edit your program **TEST** to have it check the equivalence of the commands **round(A,0)** and **int(A+.5)** using the same (a)-(e) values of exercise (7.2).

(7.4) What does the program **TEST2** do? Determine your answer by running the program for the given (a)-(d) values:

```
PROGRAM:TEST2
Prompt N,D
If D<0
   D+1 →D
int(N/10^(D-1)-10int(N/10^D)) →A
Disp A
```

(a) **N=987.654,D=2**

(b) **N=987.654,D=1**

(c) **N=987.654,D=-2**

(d) **N=987.654,D=-3**

(7.5) What do you know about **A** if the test **A=int(A)** is true?

(7.6) Use the following rand functions to produce a series of values and then describe the values you have produced:

(a) **int(10rand)**

(b) **round(rand,0)**

(c) **iPart(100rand)**

(7.7) Determine and test a **randInt** function that will produce random integers of these types:

(a) $3 \leq X \leq 7$

(b) even numbers $2 \leq X \leq 12$

(c) odd numbers $1 \leq X \leq 9$

(d) multiples of 3 between 33 and 99 inclusive

(7.8) You have a game that calls for spinning a spinner with divisions marked 1 to 10. How could you use a random function to emulate these spins?

Chapter 8.
Control Structure 3:
`For...End`

Computers are good at following instructions but not at reading your mind.
– Donald Knuth

This is my favorite control structure. It is called a counting loop because it allows you to have your program do something a given number of times. Consider how this works by entering the following simple program:

```
PROGRAM:CUBES
For(N,1,5)
  Disp N^3
End
```

When you run that program you will see it display the cubes of the integers from 1 to 5: **1, 8, 27, 64** and **125**. You can make it work harder if you change the **5** to **50**. Do so and, when you run the program, your calculator will dutifully display the cubes of numbers running from 1 to 50. However, it does this so fast that you won't see them all until it finally stops at 50. I am so simple-minded that I love to have the calculator do trivial things like that, but this control structure, like the others, is a powerful tool and has a wide range of more important uses.

You can solve that problem of the missing displays by inserting the command **Pause** (find it at **PROG CTR 8**) in your program to make it read:

```
PROGRAM:CUBES
For(N,1,50)
  Disp N∧3
  Pause
End
```

Any time your calculator comes to a **Pause** instruction, it stops processing and waits for you to tell it to go on. You do that by keying **ENTER**. Run this new program and you will see how this works. Notice as you do so that when the calculator stops to show the current display, it is still working. You can tell that by the moving line in the upper right corner of your screen. (If you leave your calculator in **Pause** and don't turn it off, you will run down your batteries.) Recall that you can break this program, if you don't want to display all 50 of those cubes, by keying **ON ENTER**.

Let's see what is going on in that program. Notice first that the **For** command line has three entries in parentheses following it. The first is a letter that will take on integer values running from the second entry to the third. This means that in the original version of **CUBES**, **N** takes on the values 1, 2, 3, 4 and 5 and for each of them performs the instructions in the loop ending at the **End**. In this case there is only one line to process, **Disp N^3**, and the effect of the loop as originally written with **For(N,1,5)** is the same as though the program read:

```
PROGRAM:COUNT
Disp 1∧3
Disp 2∧3
Disp 3∧3
```

```
Disp 4∧3
Disp 5∧3
```

It turns out that the **For...End** control structure easily processes the sum indicated by the mathematical symbol, $\sum$. Here is an example that shows how $\sum$ works:

$$\sum_{n=3}^{6} n^2 = 3^2 + 4^2 + 5^2 + 6^2 = 86$$

The left member of that equation is read: "The sum from n = 3 to 6 of n squared."

You can carry out that summation with the following program:

```
PROGRAM:SUMNSQ
0→S
For(N,3,6)
   S+N^2→S
End
Disp S
```

Notice how each time through the loop N^2 is added to the sum.

Another use of **For...End** loops is to slow down the display of information. Here is a modification of the **NAME** program designed to show how this works:

```
PROGRAM:SLOWNAME
Lbl 1
Disp "your name"
For(N,1,2000)
End
```

`Goto 1`

Recall how, without that **For...End** loop, the program ran so fast that you couldn't see the results. Even though the **For...End** loop is empty (I have included the unnecessary blank line to stress this) and doesn't accomplish anything, it still runs through the loop 2000 times and that takes some time.

The exercises will not only give you a chance to work with **For...End** loops, but will also introduce some additional features of those loops.

8.1 Exercises

(8.1) List the values of **X** that will be processed in each of these **For...End** loops:

(a) `For(X,2,6)`

(b) `For(X,231,234)`

(c) `For(X,-3,0)`

(d) `For(X,35,35)`

(8.2) Remember in primary school when you were asked to count by 2s or 5s or even to count backwards (by -1s)? Until now your **For...End** loops counted by 1, but with a third number, they allow you to count by other increments. For example, `For(J,10,30,5)` would be processed for **J** = **10**, **15**, **20**, **25** and **30**. With what values of **Y** would these loops be processed:

(a) `For(Y,5,11,2)`   (b) `For(Y,13,27,7)`

(c) `For(Y,9,5,-1)`   (d) `For(Y,3,-3,-2)`

(8.3) You want to develop a table of sin x + cos x from 0° to

$360°$ by increments of $30°$.

(a) Describe a **For...End** loop command that you could use to construct such a table.

(b) Construct and run a program that will provide you these values one at a time so you can copy them. (Hint: Include a **Pause** command.)

(8.4) When the famous mathematician Carl Friedrich Gauss was in elementary school, his teacher gave Gauss's class a "busywork" assignment: "Add the numbers from one to one hundred." Much to his teacher's astonishment, Gauss completed the exercise immediately. He paired the numbers, 1+100, 2+99, 3+98, all the way to 50+51. Each pair added to 101 and there were 50 pairs so the sum was $50 \times 101 = 5050$. In effect, what Gauss had done was show that

$$\sum_{n=1}^{100} n = 5050$$

Develop and run a program **SUMN** to check Gauss's work.

(8.5) Develop and run programs to evaluate the following sums. (You may simply edit one program to solve each part.)

(a) the sum of the cubes from 1 to 10

(b) $$\sum_{n=3}^{10} \frac{1}{n}$$

(c) $$\sum_{n=1}^{20} \sqrt{n} = \sqrt{1} + \sqrt{2} + \sqrt{3} + ... + \sqrt{18} + \sqrt{19} + \sqrt{20}$$

(8.6) Edit the program **SLOWNAME** with the **For...End** loop. You want to know about how many times through the empty loop will take one second.

(a) Change the **For(N,1,2000)** loop control to a larger number, say **For(N,1,20000)**. Estimate how many seconds long the delay is for that number.

(b) Divide 20000 by that number of seconds to get an approximate value for one second.

(c) Run the program with the **For...End** loop running from 1 to this estimate. Is your delay a reasonable approximation to one second?

  You must understand that this estimate does not take into account how long it takes to display those lines.

(8.7) Here is a complicated program with two variables, each of which is changing as the program steps are processed.

```
1 PROGRAM:FORLOOP
2 5→A
3 For(B,3,5)
4    A+B→A
5 End
6 Disp A
```

(a) Fill in the following table with the values of A and B at the end of each numbered line as the program is processed:

Line	A	B
2	5	
3	5	3
4		
3		
4		

3		
4		
6		

(b) Enter and run the program to check your final value of **A**.

(c) The final value of **B** is a bit more complicated. A **For...End** loop ends when the value of the index (that's the variable that controls it – in this case, **B**) gets too large. The value of **B** is incremented each time it completes the loop (at End) and checked against this final value. Thus, in the program **FORLOOP** the final value of **B** would be **6**, one more than the final value to be processed. Check that this is the case by entering **B** after you have run the program.

Chapter 9.
Your Calculator's
Three Screens

Programming graphics...is like
finding the square root of π
using Roman numerals.
– Henry Spencer

That section title is a bit misleading. Of course, your calculator has just one display screen. There are, however, three different output display modes for that screen and I choose to refer to them each as screens. Here then are the three screens to which I refer:

1. Your home screen. That is the one you have been using up to now.

2. Your graphing screen. As you should certainly expect from that title, this is the screen on which you draw graphs.

3. Your pixel screen. This is the screen on which you work with the 5985 individual pixels.

Your Home Screen

You are already familiar with your home screen where you perform regular calculations and display results of the programs you have created so far. In those programs you used the input commands Prompt and Input and the output commands `Disp` and `Output`.

Your Graphing Screen

Your TI-84 calculator is usually described as a graphing calculator and indeed it provides the opportunity to draw graphs. You may already have used your calculator to do so, but I will review how that is done by means of an example: Suppose you wish to graph the two equations: $y = x^3 - x + 1$ and $y = x + 1$. You can do this by following these steps, which use the keys just below the display screen:

1. Choose **FORMAT** and see that you have **AxesOn**.

2. Choose the **Y=** key and enter your functions after **Y1=** and **Y2=**.[1]

3. Choose **WINDOW** and set **Xmin -3**, **Xmax 3**, **Ymin -5**, and **Ymax 5**.

   Leave the other values all equal to **1**.

4. Key **GRAPH**.

If you followed those steps carefully, your calculator should draw those graphs as shown in Figure 9.1.

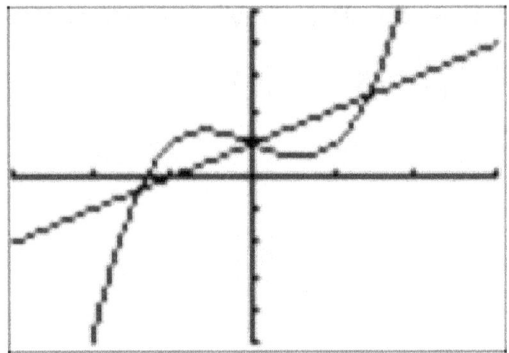

Figure 9.1 Graphs of $y = x^3 - x + 1$ and $y = x + 1$

That is your graphics screen. If you now key **CLEAR**, you will return to your home screen. Type a number on that screen, say **56**. Now key **GRAPH** again and you will return to your graph.

Notice that the graphs are not redrawn as they were in step 4; they simply remain there. Key **CLEAR** again and you will return to your home screen with its **56** still displayed. This back-and-forth activity is called toggling.[2]

You can also place points and line segments on this screen using the commands **Pt-On (** and **Line (**. (These are located in the **DRAW** menus.) You have already coordinatized your screen to display the graphs in the example. Add the point (1,-2) to your graphics screen by keying **Pt-On(1,-2)** and add the line from (-1,-3) to (2,-3) by keying **Line(-1,-3,2,-3).**

There is much more you can do with this screen, but I turn now to the screen that I find of most interest. You can do everything you can do on your graphing screen but in quite different ways. And you can do much more as well.

Your Pixel Screen

Your pixel screen, as I suggested earlier, is where you can control each of the individual pixels that make up that screen. Before I continue, I had better clarify what a pixel is. If you use your remote control device to pause a picture on your television screen and examine that screen under strong magnification, you will find that it is made up of a collection of millions of tiny colored dots. Those colored dots are pixels. Together those pixels form the images you see and, when you unfreeze the picture again, they change rapidly to show motion.

Just so on your calculator, except that on your calculator there aren't nearly as many pixels and each square pixel is either dark (on) or light (off). Your calculator pixels are also big enough that, if you look closely, you can see them. It is important that you understand that everything you see on that screen is displayed as collections of those pixels, some on, some off. If all of them are on, you see a black screen; if all of them are off, you see a blank screen.[3] It is easy to confuse the points on your graphics screen with the pixels on your pixel screen. The graphing screen does make use of pixels, but it translates them into the coordinate system that it sets up with **WINDOW**. The pixel screen uses the basic coordinates assigned by your calculator.

Pixel coordinates are named (row,column) just as the coordinates on your home screen were named. But while your home screen had only 128 locations, your pixel screen has many more pixels you can control. Here is the pixel screen with the limiting pixels and a few others identified:

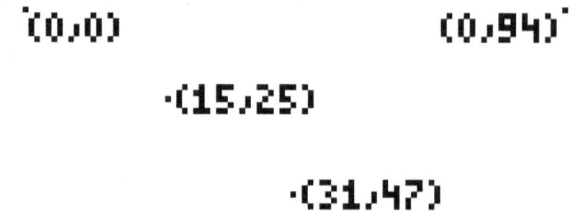

Figure 9.2 Pixel coordinates

There are 63 rows on this screen numbered from 0 to 62, and 95 columns numbered from 0 to 94. It is the associated arithmetic (63 × 95) that gives you the 5985 total number of those screen locations. You can turn on any one of those pixels with the command **Pxl-On(** and you can turn off a particular pixel with **Pxl-Off(.** For example, the command **Pxl-On(10,10)** would turn on a pixel in the upper left part of your screen.

There is no **Line(** command as there was with your graphing screen. One way to draw a line on your pixel screen is to use a

For...End loop. Here is a program that will draw a line from pixel (20,20) to pixel (50,50):

```
PROGRAM:LINEEX
AxesOFf
FnOff
ClrDraw
For(A,20,50)
   A→B
   PixelOn(A,B)
End
```

It is useful to include those first three lines in any program involving the pixel screen. If you omit them, you will often pick up unwanted graphs or other data from earlier activity. **ClrDraw** is located in the **DRAW** menu and **AxesOff** and **FnOff** may be found in the **CATALOG** menu. **CATALOG** is useful for finding any command: once you have accessed the menu, key a letter (no **ALPHA** necessary) to send you to commands beginning with that letter. Then scroll to the one you want. **FnOff**, for function off, turns off the display of any graphs you have assigned in the **Y=** menu. You must then remember the next time you want to draw graphs on your graphing screen to key **FnOn**. As an alternative to using **FnOff**, you can simply clear all the **Y=** graph functions before you program your pixel screen.

You can also print on this screen and, because the letters are smaller, you can fit in more of them. Using the command **Text(** (found by scrolling up in **DRAW**), you can place letters, words or phrases on your pixel screen using this command. I offer a final revision of your much-abused program CTOF to show how this can be done:

```
PROGRAM:CTOFLAST
AxesOFf
FnOff
ClrDraw
Input "CELSIUS TEMP? ",C
9/5*C+32→F
Text(10,10,C," DEGREES CELSIUS")
Text(25,15,"IS EQUIVALENT TO")
Text(40,5,F," DEGREES FAHRENHEIT")
```

That program, with **C=37**, will produce the screen:

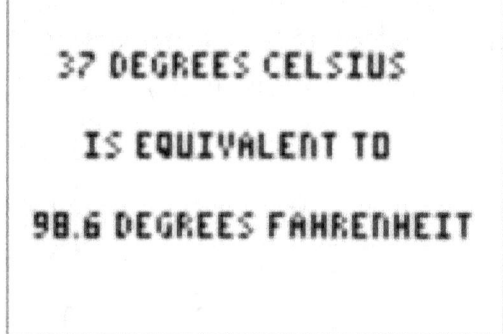

Figure 9.3 Your CTOF Screen

Notice how with the Text command you can place data (**C** and **F**) and text together in the same statement. You could not do that with the **Output** command. This feature allows you to save three lines from the earlier **CTOF** version and improve the format of the output as well.

- - - - -

1 Because **X** occurs so often in programs, there is a shortcut: the single **X,T,θ,n** key substitutes for **ALPHA X**.

2 The "caps lock" key on your computer keyboard is another toggle

key; you change to capital letters when you press it and back to lower case when you press it again.

3 As exercise (9.4) will show, your active screens do not take up the full display panel of your calculator.

9.1 Exercises

(9.1) Turn axes on and graph each of the following, using the window limits indicated:

(a) $y = x^2$, Xmin=-2, Xmax=2, Ymin=-.2, Ymax=3

(b) $y = (\sin x)^2 + \cos x$, Xmin=0, Xmax=360, Xscl=30, Ymin=-1.2, Ymax=1.5 (Be sure your calculator is in Degree mode.) Note that mathematicians would display this function as $y = \sin^2 x + \cos x$.

(c) $y = \ln x$ and $y = -\ln x$, Xmin=0, Xmax=5, Ymin=-2, Ymax=2

(9.2) A problem that many calculator users have is locating a graph appropriately on the screen. One way to resolve this problem is to begin with a large window, using coordinates like -20 ≤ X ≤ 20 and -20 ≤ Y ≤ 20, and then narrow that window as appropriate.

Use that technique to draw the following graphs and indicate the X and Y limits you chose:

(a) $y = 3x^3 - 5x^2 + 12$ (b) $y = x + 1/(2x)$

(c) $y = x - 3 \ln x$ (d) $y = 2.5e^{-.5(x-3)^2}$

(e) $y = (\sin x)^2 - \cos x$. This time increase your X-limits to 0 ≤ X ≤ 360 and change the X-scale to 30.

(9.3) When you use different scales for x and y in graphing, you get figures like circles distorted. You can see this by using the

`Circle(` function that appears in **DRAW**. If you key `Circle(x,y,r)` you will plot a circle with center at (x,y) and radius r in your graphing window. The following exercises will address this distortion problem, but first, clear your graphs in **Y=**.

(a) Change your **WINDOW** values to $-1 \leq x \leq 4$ and $-1 \leq y \leq 4$ and in your home screen enter `Circle(2,2,1)`. Clearly, your resulting figure is not a circle. What is it?

(b) The problem here arises from what is called the aspect ratio of your screen. If it was a perfect square and the x and y scales were the same, you would have created a circle. Your screen instead is (in terms of pixels) 95 wide by 63 high. That is very close to the aspect ratio of 3:2 for x:y. You can achieve this ratio by setting your ranges in WINDOW in proportion to 3:2. Change your window values now to $-1 \leq x \leq 5$ and $0 \leq y \leq 4$. Now plot `Circle(2,2,1)`. Did it produce a circle? What were the lengths of the x and y ranges?

(c) Although you cannot use the function `Circle(` on your pixel screen, distortion can occur there as well, and there you don't have **WINDOW** to assist you. The pixels are squares, however, so that should help. Try the program **BIGSQ** to see if you get a square:

```
PROGRAM:BIGSQ
AxesOFf
FnOff
ClrDraw
For(A,10,50)
  Pxl-On(A,10)
  Pxl-On(A,50)
```

```
  Pxl-On(10,A)
  Pxl-On(50,A)
End
```

What are the first ten pixels that are turned on by this program? Are some of them duplicates?

(9.4) On page 54 I claimed that, if you turned on all your pixels, you would have a black screen. I had better clarify that here. The program **DARK** will turn on all your pixels.

```
PROGRAM:DARK
AxesOFf
FnOff
For(R,0,62)
  For(C,0,94)
    PxlOn(R,C)
    Pxl-On(A,10)
  End
End
```

(a) Run that program and watch the pixels being turned on. I omitted the **ClrDraw** command from the program because any left-over dark pixels on your screen will stay dark as though they are being swept under by the unfolding rug of program-produced pixels. That program darkens the active part of your calculator's display. The remainder of your screen will never be active. Does this program turn on pixels running across the screen or down it?

(b) The structure of this program is important. That is why I indented it. (This is standard programming practice for clarification and I will continue to do this in future programs.) Remember the **GRADE** program back on page 16. It had an **If...Then...Else...End** structure embedded in the **Else** part of another **If...Then...Else...End**. Here we have one **For...End** loop embedded in another **For...End** loop.

When your calculator reaches the **For(R,0,62)** it sets **R** equal to the first value, **0**. It then goes to the **For(C,0,94)** line and sets **C** to its first value, **0** also. In the next line **Pxl-On(R,C)** it sets the pixel **(0,0)** on. Now it reaches the first End and turns back to get its next instruction. This is the **End** for the (inner) **For(C,0,94)** loop so it goes back to that command and takes the next value of **C**, which will be **1**. (Remember, it is counting up to **94**.) Since **R** remains **0** and **C** is now **1**, the **Pxl-On(R,C)** command turns on the point **(0,1)**. That's row **0**, column **1**. That inner loop will continue to repeat for pixels **(0,2)**, **(0,3)**, and so on until it gets to pixel **(0,94)**. After it turns on that pixel it jumps out of that inner loop but immediately hits the **End** for the outer loop. Now **R** increments from **0** to **1**, and again begins processing that inner loop, darkening **(1,0)**, **(1,1)**, and so on.

What will be the final five pixel values darkened?

(c) Edit the program by placing the command **For(C,0,94)** ahead of the line **For(R,0,62)** instead of after it. Guess what will be the effect of this change before you run the program again. This time which way are the pixels darkened?

(9.5) The following program will give you information about

how letters and numbers may be assigned to and appear on your pixel screen.

```
PROGRAM:TEXTSIZE
AxesOff
FnOff
ClrDraw
Text(0,0,"ABCDEFGHIJKLMNOPQRSTUVWXYZ")
Pxl-On(0,3)
Text(6,0,"B")
Text(12,0,"C")
Text(18,0,"D")
Text(24,0,"E")
Text(30,0,"G")
Text(36,0,"H")
Text(42,0,"I")
Text(48,0,"J")
Text(54,0,"K")
Text(10,10,"XYZ",123456789)
Text(20,10,"SPACE KEY:1 PIXEL WIDE")
Text(30,10,"TWO  SPACES  IS  BETTER")
Text(40,10,"THREE   IS   OKAY   TOO")
Text(57,91,"O")
```

Here is what that would display on your pixel screen:

Figure 9.4 Program **TEXTSIZE** Display

Today we are used to letters and numbers being represented by square dots. Every electronic device uses that kind of display. It is worth giving some thought to your calculator's screen to gain some insights into not only that screen but character representation in general. Enter and run the program **TEXTSIZE** and look carefully at your screen as well as the program commands to answer the following:

(a) Notice that you placed the letters **ABC...** beginning at the point **(0,0)**. The pixel your program placed at the point **(0,3)**, however, shows that your letters leave a blank line above them. Here is the way the letters A and B are printed with pixels:

Figure 9.5 Pixels for **A** and **B** when printed on Pixel and Home screens

When you look at them like this, they don't look like much. (On the pixel screen, they look to me like an **H** and an **E** with a couple of dots thrown in.) When they appear small on your screen, however, your eyes help you to read them. The letter **N** seems to be a problem to represent. Why is this?

(b) In what small way do the **H**, **M**, **U** and **W** differ on the pixel screen?

(c) Compare the pixel screen **N** with the home screen **N**.

(d) Compare the number of lines and columns you can print on the pixel screen and the home screen.

Chapter 10.
Control Structures 4:
`While...End`
and 5:
`Repeat...End`

Always code as if the guy
who ends up maintaining your code
will be a violent psychopath
who knows where you live.
— Martin Golding

I am including these final two major control structures together because, although they have important differences, in many cases they are interchangeable.

It will be helpful if you think of these control structures as larger informal phrases: **While** this test is true, do the following steps over and over. And **Repeat** the following steps over and over UNTIL this test is true. (Many people think of **Repeat** as **Repeat until**.)

Here are non-calculator examples of how these loops work:

```
While weather is good
   Head for the beach
   Use suntan lotion
End
```

   and

```
Repeat until weather turns bad
  Head for the beach
  Use suntan oil
  End
```

Here are two simple and quite similar mathematical examples:

```
PROGRAM:WHILLOOP
Prompt N
While N≤5
  Disp N
  N+1→N
End
```

```
PROGRAM:REPLOOP
Prompt N
Repeat N>5
  Disp N
  N+1→N
End
```

There are several important points to be made about those programs:

1. Those loops accomplish something that could also be done with a **For...End** loop:

```
PROGRAM:FORLOOP2
Prompt N
For(I,N,5)
  Disp I
End
```

Notice that the **While** and **Repeat** loops require you to increment **N** with the command **N+1→N**, whereas the

For...End loop does that for you.

2. All three programs display the same values when integers less than 6 are entered. For example, if the initial value is **N=3**, the output for each would display **3**, **4** and **5**.

3. The output of the **REPLOOP** program differs from the others if the initial value of **N>5**. The **WHILLOOP** and **FORLOOP2** programs would display no output numbers but **REPLOOP** would display **N**. This is because the test for a **Repeat** loop is done at the end of the loop rather than at the beginning. Thus all **Repeat** loops are processed at least once.

It is important that you understand, moreover, that the **While...End** and **Repeat...End** control structures are not simply complicated versions of **For...End**. They address situations in which **For...End** would not suffice. For example, suppose we want to develop a program that will find the average (or mean) of a set of positive data, when we do not know how many items we will include. Here is a program that will do this for you:

```
PROGRAM:MEAN
0→S
0→N
Prompt D
While D>0
   S+D→S
   N+1→N
   Prompt D
End
S/N→M
Disp M
```

That program accumulates the sum of the values you enter in **S** and keeps track of how many values with **N**. When you have finished entering data, you can enter zero or any negative number to jump out of the **While...End** loop. At the end you divide the sum of the items by the number of items to produce the average. In the next section I will show you how you can keep track of what is going on in this kind of program step by step.

10.1 Exercises

(10.1) Here is a simple program that uses the **While...End** control structure. It displays the values **8**, **9** and **10**:

```
PROGRAM:EXWH101
8→N
While N≤10
  Disp N
  N+1→N
End
```

(a) Write and test a program **EXREP101** that accomplishes the same thing with a **Repeat...End** control structure.

(b) Write and test a program **EXFOR101** that accomplishes the same thing with a **For...End** control structure.

(10.2) Here is a program that will find the range of a set of positive data, that is, the highest and lowest values:

```
PROGRAM:RANGE
10^99→L
0→H
Prompt D
```

```
While D>0
   If D<L
      D→L
   If D>H
      D→H
   Prompt D
End
Disp L,H
```

   (a) Give two examples of numbers that you can enter for **D** to
       end this **While...End** loop.

   (b) Notice how the low value, **L**, is set very high in the first
       line of the program. This seems contrary to what you would
       expect, but your very first entered **D** will be less than this
       and replace it in step 7. How does this work for the high
       value, **H**.

(10.3) There are two Prompts in the programs **MEAN** and
       RANGE. Here is the program **MEAN** compared with a bad
       alternative:

```
PROGRAM:MEAN
0→S
0→N
Prompt D
While D>0
   S+D→S
   N+1→N
   Prompt D
End
S/N→M
Disp M
```

```
PROGRAM:BADMEAN
0→S
0→N
While D>0
   S+D→S
   N+1→N
   Prompt D
End
S/N→M
Disp M
```

(a) The problem arises in **BADMEAN** when you first arrive at
the third program line. Do you have a value of **D** to test at
this point? (Your calculator may have one, but is that the
one you want here?)

(b) It turns out that **BADMEAN** can easily be converted into a
correct program using **Repeat...End** instead of
While...End. You need only adjust line 4 to accomplish
this. Make that adjustment and test the program. What
should be the new line 4?

(10.4) Another problem comes up in some programs caused by
the fact that a **Repeat...End** loop is always processed
once. Notice what happens with the following minor
modification of the **REPLOOP** program when you enter a
particular value:

```
PROGRAM:REPLOOP2
Prompt N
Repeat N≠5
   Disp N
```

```
    N+1→N
End
```

(a) What values are output with **N** initiated with 3 and 4?

(b) What happens when **N** is initiated with 5? Why?

Chapter 11.

Subprograms and Tracing

There are two ways
to write error-free programs;
only the third one works.
– Alan Perlis

In each of the programs that has involved the pixel screen, we have included lines that prevent axes and graphs from appearing and that cleared the screen. Writing those lines over and over is time consuming so why not find a way to include them in a single step? Aha! We can do this with a subprogram. This will seem a bit complicated at first, but you will soon get used to this technique and, when you develop larger programs, subprograms do save much typing. This example will show you how to proceed:

Develop a new program **PXLSETUP**:

```
PROGRAM:PXLSETUP
AxesOff
FnOff
ClrDraw
```

Now, when you want to write a program using your pixel screen, you need only enter that program name, **PXLSETUP**. You do not, however, simply type those letters. You have to call up that program.

Of course, you will want to use this subprogram in developing new programs, but here we'll edit a program you have already developed, **BIGSQ**. Call up **BIGSQ** and **CLEAR** the first

63

program line. Replace it by keying **PRGM**, scroll right to **EXEC** and down to **PXLSETUP**. When you press **ENTER**, the command **prgmPXLSETUP** will appear on that first line of your program. Now you don't need those setting-up lines in the program and you can delete them to have:

```
PROGRAM:BIGSQ
prgmPXLSETUP
For(A,10,50)
   and so on....
```

When you run that program, you will achieve the same results as you did before. What happens is that when it comes to that line **prgmPXLSETUP**, your calculator transfers control to that program. It then runs through the commands of that program and when finished it returns to your original program to proceed.

There is a command **Return** that you can use within such subprograms. It is not necessary, however, to include this command at the end of a subprogram.

Yes, subprograms can save steps, but they can also save editing time. Suppose we have a program that includes a function line like **5X+7→Y**. That same function might even appear in several places in the program. The program might appear like this:

```
PROGRAM:EXAMPLE
<some program lines>
5x+7→y
<more program lines>
5x+7→y
<still more program lines>
```

Now you want to make your program EXAMPLE work for other functions, like $X^2-13\rightarrow Y$.

You can, of course, go back to your program and change each of those lines, but you can also create a one line program **FNXTOY** with a function in it. You could start with the original function, **5X+7→Y**, so the program would appear as:

```
PROGRAM:FNXTOY
5X+7→Y
```

Then go back to the program **EXAMPLE** and replace those **5X+7→Y** lines with **prgmFNXTOY** so you have:

```
PROGRAM:EXAMPLE
<some program lines>
prgmFNXTOY
<more program lines>
prgmFNXTOY
<still more program lines>
```

Now your **EXAMPLE** program is more general and you can change the function to anything you wish by modifying that one step in **FNXTOY**.

Tracing Programs

You have been working with very brief programs so far, but they rapidly get more complex and you need a way to check how your programs are working. The ability to trace programs will help you both in constructing them and in finding and correcting errors.

What you want to do in tracing a program is make a table of values, preferably next to a printed copy of your program. I will show you what I mean by tracing the program MEAN from the last section. I will trace the program for the input values (of **D**) of 80, 90, 100 and 0, the last to escape the **While...For** loop:

PROGRAM:MEAN

1   0→S
2   0→N
3   Prompt D
4   While D>0
5   S+D→S
6   N+1→N
7   Prompt D
8   End
9   S/N→M
10 Disp M

Line	S	N	D	M
1	0			
2	0	0		
3	0	0	80	
5	80	0	80	
6	80	1	80	
7	80	1	90	
5	170	1	90	
6	170	2	90	
7	170	2	100	
5	270	2	100	
6	270	3	100	

7	270	3	0	
9	270	3	0	90

The table only keeps track of variables. Notice that there is no entry for the important controlling lines 4 and 9, but those must be checked in the process.

No one likes to write more than is necessary and you will soon find yourself recording only the changes as you trace programs and you will omit program line numbers as well. (After all, they don't appear in your calculator.) Then your trace would look like this:

S	N	D	M
0	0	80	
80	1	90	
170	2	100	
270	3	0	90

I urge you, however, to begin making more complete traces like that of the first table.

11.1 Exercises

(11.1) Here are a program and subprogram. What value of D will be displayed when program A is run?

```
PROGRAM:A
5→B
progSUBC
Disp D
```

```
PROGRAM:SUBC
If B<7
Then
   8→D
   Return
End
Disp D
```

(11.2) What value of **D** would be displayed in exercise (11.1) if the **Return** were omitted from subprogram **SUBC**?

(11.3) To practice making the kind of changes that occur in program steps, copy and complete the following table, assigning new values according to the given instructions:

command	A	B	C	D
initial values	1	2	3	
A+C→D				
$C^2 + D^2 \rightarrow A$				
$\sqrt{A} \rightarrow B$				
(A + B + C) / C → D				
$\text{int}\left(\sqrt{B}\right) \rightarrow C$				
If A < 7 : Then : 3 → D				

(11.4) Ordering numbers with your calculator is a more complex task than you might expect. Trace the values of **A**, **B** and **T**, a temporary value introduced to help the process, for the following program. First use the values A=3 and B=5, then the values A=5 and B=3.

```
PROGRAM:LOHI
Prompt A,B
If A>B
Then
    A→T
    B→A
    T→B
End
Disp A,B
```

(11.5) Here is the program that most beginners think would do just as well. Trace it with **5** for **A** and **3** for **B** to see why it doesn't work.

```
PROGRAM:BADLOHI
Prompt A,B
If A>B
Then
    B→A
    A→B
End
Disp A,B
```

(11.6) How would you modify the program **LOHI** to order the values high then low?

(11.7) Ordering three values is even more complicated. Trace the following program for the values:

(a) 1,3,2   (b) 3,1,2   (c) 3,2,1

```
PROGRAM:LOHI3
Prompt A,B,C
If A>B
Then
```

```
        A→T
        B→A
        T→B
End
If B>C
Then
        B→T
        C→B
        T→C
End
If A>B
Then
        A→T
        B→A
        T→B
End
Disp A,B,C
```

(11.8) In the program **LOHI3**, the right column is the same as part of the left column. Why is it necessary to repeat those lines?

Sorting is an important task for a calculator or computer and computer scientists have devised a variety of techniques to carry out this process for longer lists of numbers. You will find that you can sort the **LIST**s you will meet in Section 12 with a single command, but much is being done behind the scenes when you use that command.

Chapter 12.

Matrices

Should array indices start at 0 or 1?
My compromise of 0.5 was rejected
without, I thought,
proper consideration.
– Stan Kelly-Bootle

Matrices are very important mathematical objects. They play a central role in sophomore or junior university courses like Linear Algebra or Vector Calculus. But don't let that scare you. Matrices are simply arrays of numbers. For example, a baseball scoreboard is a matrix. For a regular nine-inning game, omitting the summary columns for runs, hits and errors, the numbers might look like this:

$$
\begin{pmatrix}
1 & 2 & 3 & 4 & 5 & 6 & 7 & 8 & 9 \\
0 & 0 & 0 & 1 & 0 & 0 & 3 & 1 & 0 \\
1 & 0 & 0 & 0 & 0 & 0 & 0 & 1 & 4
\end{pmatrix}
$$

It should be evident that matrices are made up of rows and columns. In this example, the numbers in the first row represent the innings, the second row the runs scored by the visiting team and the third row the runs scored by the home team in those innings.

Matrix elements, the individual entries in the matrix, are indicated by row and column in that order.[1] Here, for example, is the form of a 3 × 4 matrix as it would be represented mathematically:

$$\begin{pmatrix} a_{11} & a_{12} & a_{13} & a_{14} \\ a_{21} & a_{22} & a_{23} & a_{24} \\ a_{31} & a_{32} & a_{33} & a_{34} \end{pmatrix}$$

The calculator uses only slightly different notation to convey this same information:

$$\begin{pmatrix} A[1,1] & A[1,2] & A[1,3] & A[1,4] \\ A[2,1] & A[2,2] & A[2,3] & A[2,4] \\ A[3,1] & A[3,2] & A[3,3] & A[3,4] \end{pmatrix}$$

Your TI-84 calculator provides access to ten different matrices, labeled **[A]** to **[J]**. Each of those matrices allows 99 rows and 99 columns, but don't count on using such big arrays. You would run out of memory long before you created a 99 by 99 matrix. A 50 × 50 matrix filled with zeros uses over 22 kilobytes of memory. That doesn't mean that you cannot make use of all those rows or columns: you could, for example, accommodate a very long extra-inning baseball game with a 3 × 99 matrix. A more realistic example would be three months of temperature lows and highs in a 3 × 90 or 90 × 3 matrix.

I can best show you how to work with matrices in programming by means of an example. Here is a program that sets up (initializes) a 3 × 4 matrix filled with zeros and then replaces those zeros by entering in each element the sum of its row and column.

```
PROGRAM:MATRIXEX
1   {3,4}→dim([A])
```

```
2  Fill(0,[A])
3  For(R,1,3)
4    For(C,1,4)
5      R+C→[A](R,C)
6    End
7  End
```

After running that program, your calculator's matrix **[A]** would be:

$$[A] = \begin{pmatrix} 2 & 3 & 4 & 5 \\ 3 & 4 & 5 & 6 \\ 4 & 5 & 6 & 7 \end{pmatrix}$$

Now let's see what is happening in that program:

1. In line 1 the matrix **[A]** is established and its dimensions set. When entering this line, key the values up to the →. The function **dim** is found in **MATRIX MATH** and then you must access the matrix name **[A]** in **MATRIX** as well. (You cannot simply type **[A]**.)

2. In line **2** your matrix is filled with **0**s. The command **Fill** is in **MATRIX MATH**.

3. Lines **3** through **7** are nested **For...End** loops. The outside loop sets **R = 1** and the inside loop is run with that value. Thus the **(R,C)** values are run **(1,1),(1,2),(1,3),(1,4)**. Once that is done, the outer loop increments to **R = 2** and the inner loop is run again, producing **(2,1),(2,2),(2,3),(2,4)**. Again the outer loop is incremented and you produce **(3,1),(3,2),(3,3),(3,4)**.

4. As each of those matrix elements is accessed the program replaces the zero already there with the sum of its row and column.

5. You can see your resulting matrix by keying **MATRIX [A] ENTER ENTER**

Matrices have many uses. They can be operated on mathematically: added and multiplied, for example. Those uses require some additional mathematics that I choose not to introduce here. What then are matrices good for? At the simplest level they provide additional storage. Your calculator provides 27 keyboard storage locations, **A** through **Z** and **Θ**. A matrix can provide many additional storage locations and they are systematically organized for ready access.

A final definition: A *vector* is a single row or single column matrix. You probably have thought of a vector as a kind of arrow and it is that as well. But consider the vector from the origin pointing at the point (5,3). You can represent that arrow as the 1 × 2 row matrix **[A] (1,1)=5** and **[A] (1,2)=3** or the 2 × 1 column matrix **[A] (1,1)=5** and **[A] (2,1)=3**. You should see from this that you can increase your vector dimensions to 3-D and even beyond.

- - - - -

1 Unfortunately, there is an inconsistency here with standard graph coordinates. Graph coordinates like (x,y) are listed in the order x-column, y-row. Pixel coordinates, on the other hand, follow the form of matrices as they are listed in row-column order.

12.1 Exercises

(12.1) Develop and test a program that initializes this matrix. You can use statements of the form **9→[B] (1,2)**:

$$[B] = \begin{pmatrix} 7 & 9 \\ 1 & 5 \end{pmatrix}$$

(12.2) Develop and test a program that uses a **For...End** control structure to initialize this vector:

$$[C] = \begin{pmatrix} 1 & 2 & 3 & 4 & 5 & 6 & 7 & 8 & 9 \end{pmatrix}$$

(12.3) You can, of course, fill individual matrix elements with statements like those you used in exercise (12.1), but it is always useful to do so efficiently. Develop and test a program to initialize and fill the following matrix:

$$[D] = \begin{pmatrix} 1 & 2 & 3 & 4 & 5 & 6 & 7 & 8 & 9 \\ 7 & 14 & 21 & 28 & 35 & 42 & 49 & 56 & 63 \\ 45 & 40 & 35 & 30 & 25 & 20 & 15 & 10 & 5 \end{pmatrix}$$

(12.4) Once you have constituted a matrix, you can draw upon it in another program.

(a) Develop and test a vector **[E]** whose coordinates are the square roots of the numbers from 1 to 10. Your vector will display the decimal values of these square roots.

(b) Develop and test a two line program that prompts for an input number between one and ten and uses the vector you created in exercise (12.4) to display the square root of that number.

Chapter 13.

Lists and Strings

If debugging is the process
of removing bugs,
then programming must be the process
of putting them in.
– Edsger Dijkstra

Unlike matrices, lists and strings are not mathematical structures. They serve, however, as useful locations in which to store and manipulate data. I will consider them separately.

Lists

You can store up to 999 numbers in a list. You can work with lists on your Home screen or in programs. You might, for example, wish to store the first five prime numbers[1] in list L_1. To do this you could simply enter $\{2,3,5,7,11\} \rightarrow L_1$. Once you have done that, you can refer to individual primes. For example, enter $L_1(4)$ to retrieve the fourth prime, 7, from your list.

Lists can be modified in a variety of ways:

1. You can copy one list to another with a command like $L_1 \rightarrow L_2$.

2. You can append additional numbers to a list. You could, for example, add a sixth prime to the list L_1 you created in the last paragraph by entering $13 \rightarrow L_1(6)$

3. You can sort your lists with **SortA** for ascending (small to large) and **SortD** for descending (large to small). (Note:

these and other list functions are found under **LIST**.) If you wanted L_1 sorted from large to small, you would simply enter **SortD(L_1)**. The result would be the list $\{13,11,7,5,3,2\}$.

4. You can use dim in several ways. You can use it to create a 3-item list L_4 with **3→dim(L_4)**. All of its items will be **0**. You can use **dim** to determine how many items there are in list L_1 with **dim(L_1)**. And you can shorten or lengthen a list with it. For example, you could enter **4→dim(L_1)** to reduce the number of primes listed to four. If you lengthen a list using **dim**, it will fill the extra elements with zeros.

5. You can use **Fill** to fill a list, just as you did with matrices.

6. If two lists have the same dimension, you can add, subtract, multiply or divide them. For example, if you enter $\{1,2,3\}$→L_2 and $\{4,5,6\}$→L_3, then the instruction L_2/L_3→L_4 would create L_4 as $\{.25,.4,.5\}$.

7. Finally, you can concatenate two lists (that odd term means to append one list at the end of another) with the function augment. For example, using the lists defined in (6), the command **augment(L_2,L_3)→L_5** would make $L_5 = \{1\ 2\ 3\ 4\ 5\ 6\}$.

Strings

Whereas lists store numbers, strings store letters and numbers, but an individual string will contain exactly one entry. For example, if you enter **"HELLO R2D2."→Str1**, you can call up that string by simply entering **Str1**. (Note that you find a list of ten string names with **VARS String**.)

-----

1 Recall that prime numbers are positive integers that have exactly two

divisors, themselves and one.

13.1 Exercises

(13.1) Enter $2 \rightarrow A$ and then enter $\{A, Z\} \rightarrow L_6$.

   (a) When you now enter L_6, what string results?

   (b) Where did those numbers come from?

(13.2) Enter the following: $\{1, 2, 3, 4\} \rightarrow L_1$ and $\{4, 3, 2, 1\} \rightarrow L_2$, then first guess what will result from the following, then check your answers with your calculator:

   (a) $3L_1$ (b) $L_1 + L_2$ (c) L_1^2 (d) $L_1 \times L_2$

   (e) `augment(`L_2, L_1`)`   (f) `sortD(`L_1`)`

   It is important to notice that only part (f) changes a list: it changes L_1.

(13.3) Guess what this program accomplishes, then enter and run it to check your answer:

```
PROGRAM:EX133
10→dim(L₅)
For(N,1,10)
  N²→L₅(N)
End
Disp L₅
```

(13.4) Store your first name in **Str2**.

   (a) Write a program **EX134A** that prints **HELLO** followed by your first name near the middle of your home screen.

   (b) Write a program **EX134B** that prints **HELLO** followed by your name near the middle of your pixel screen.

   (c) Enter **"HELLO "**$\rightarrow$**Str3**. Then use **Str3+Str2** to print **HELLO** followed by your name near the middle of your home screen.

(d) Use **Str3+Str2** to print **HELLO** followed by your name near the middle of your pixel screen.

(13.5) There are times when you wish to extract one or more letters from a string. If **Str3** is **"HELLO "** The command **sub(Str3,2,3)** would produce the string **"ELL"**. The **2** in that command names the initial position in **Str3** and the **3** represents the number of characters to be selected for the subsstring. Use **CATALOG** to find **sub(**.

(a) What command would pick out the single letter **"O"** from Str3?

(b) Develop and run a program using **Str3**, **sub(** and a **For...End** loop that prints out on successive lines:

    H
    HE
    HEL
    HELL
    HELLO

Chapter 14.

Putting what you learned to use

*Mathematicians stand
on each others' shoulders
and computer scientists stand
on each others' toes.*
– Richard Hamming

You have now met the major programming tools that are useful in addressing problems with your calculator. In introducing you to those tools, I used simple, indeed trivial, examples. Now you are invited to use those tools to address more significant problems. I will provide assistance, but the more you can work on these problems on your own, the better.

Before setting you to work, however, I offer an example related to the history of computation.

Duplation and Mediation

Somewhere along the line you must have been introduced to Roman numerals, numerals that use the symbols I, V, X, L, C, D, and M to represent numbers. A question arises. Surely people in those early times needed to multiply. We know how to multiply 25 × 21, but how did the Romans multiply the corresponding XXV × XXI without our more efficient Hindu-Arabic numeration system? Almost certainly they used a very different multiplication method that dated from at least 1650 BCE.[1] The method is called duplation and mediation (doubling and halving) or Russian peasant multiplication. All their method (aka algorithm) required was knowing how to double numbers, how to take half (not bothering with remainders) and how to add and subtract.

Those tasks they could accomplish easily, even with Roman Numerals. I will explore their method instead using our numeration. Here are the steps you can follow to carry out multiplication of positive integers to obtain their product:

1. First write down the two factors and in parallel columns write the halves (discarding remainders) of the numbers in the second column down to 1 and, to their left, the same number of doubles in the first column. Your work will then appear like this:

$$\begin{array}{ccc} 25 & \times & 21 \\ 50 & & 10 \\ 100 & & 5 \\ 200 & & 2 \\ 400 & & 1 \end{array}$$

2. Next cross out the even numbers in the right column and the numbers in the left column across from them.

$$\begin{array}{ccc} 25 & \times & 21 \\ \cancel{50} & & \cancel{10} \\ 100 & & 5 \\ \cancel{200} & & \cancel{2} \\ 400 & & 1 \end{array}$$

3. Your product is then the sum of the remaining numbers in the left column, in this case 525.

The mathematical justification for this process depends on what we today would record in algebraic form as $xy = x(y-1) + x$ and $xy = (2x)(y/2)$. Here is the exercise worked out in this way.

$$25 \times 21 = 25(20 + 1)$$
$$= 25 \times 20 + 25$$
$$= 50 \times 10 + 25$$
$$= 100 \times 5 + 25$$
$$= 100\,(4 + 1) + 25$$
$$= 100 \times 4 + 100 + 25$$
$$= 200 \times 2 + 100 + 25$$
$$= 400 \times 1 + 100 + 25$$
$$= 525$$

There are just two rules operating here:

1. Whenever the second factor is even, replace the product xy by $2x \times y/2$. This occurs between the second and third, third and fourth, sixth and seventh and seventh and eighth lines.

2. Whenever the second factor is odd, separate one first factor to include in the sum, always leaving the remaining product even. This is where the identity $xy = x(y - 1) + x$ applies. This occurs between the first and second and fourth and fifth lines.

I will use those two rules to develop a program that multiplies positive integers by this method:

```
    PROGRAM:DUPMED
1   Prompt X,Y
2   0→P
3   While Y≥1
4     While Y/2=int(Y/2)
5        Y/2→Y
6        2X→X
7     End
```

```
 8      Y-1→Y
 9      P+X→P
10   End
11   Disp P
```

If you run that program, you will find that it not only multiplies numbers like 25 × 21 but also numbers like 29778 × 99979. I hope that it will be useful to you to follow how I developed that program. Here is how I thought it through:

1. Line **1** of the program is straightforward. I had to acquire the factors to be multiplied.

2. Now I wanted to program algorithm steps 1 and 2. How would I accomplish step 1? I needed to decide if **Y** is even. I can do that with the test **X/2=int(X/2)**. I then thought of writing:

```
If X/2=int(X/2)
Then
   Y/2→Y
   2X→X
End
```

However, I realized that this sometimes happens several times in a row, as in lines three and four and seven and eight of the example. For that reason, I changed from the **If...Then...End** structure to a **While...End** structure. That gave me lines **4** through **7** of the program.

3. Next I needed to carry out step 2 of the algorithm. When **Y** is odd, it is reduced by one (to make it even and at the same time the value of **X** is added to the product. That gave me steps **8** and 9 of the program. I also had to go back and initialize **P** as **0** in line **2**.

4. Those steps completed the algorithm, but they had to be repeated until the process is finished. I noticed that the **Y** column in the example given in the text reduced to **1**. This gave me the basis for the outer **While...End** loop governed by lines **3** and **10** that repeated the process until that happened.

5. Finally, step **11** displays the resulting product that was being accumulated in line **9**.

I have taken you through that complex process as it relates to an interesting algorithm. In the following exercises you will have a chance to try your own hand at developing similar programs.

- - - - -

1 That is the date of the British Museum's Rhind papyrus which includes a form of this method.

14.1 Exercises

(14.1) Trace the program **DUPMED** with **X** = **25** and $Y = 21$.

(14.2) The current population of the USA is about 227 million and the growth rate is approximately 0.7%. The current population of Mexico is approximately 70 million with a growth rate of 3.3%. If you assume that rate of growth will continue, in how many years will the population of Mexico exceed that of the USA? This is an obvious candidate for a **While U>M...End** loop, first setting **227→U**, **70→M** and the current year to **Y**.

(14.3) Develop a program that will find the perimeter and area of any regular polygon, given its number of sides, **N**, and the length of its radius, **R**. (A radius of a regular polygon is a segment from the center of the polygon to any vertex; an apothem is a segment from the center perpendicular to a side of the polygon.)

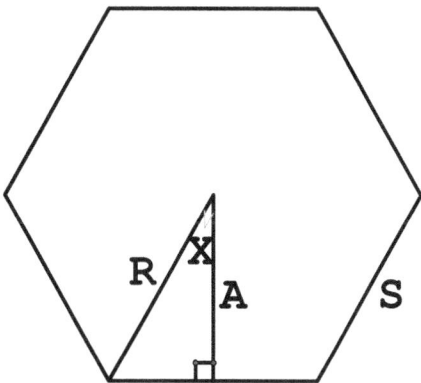

Figure 14.1 A regular hexagon with radius, **R**, and apothem, **A**, indicated

(a) Justify for yourself that the following formulas (with P for perimeter and T for area) apply to all regular polygons:

$$X=180/N \quad S=2Rsin(X)$$

$$A=Rcos(X) \quad P=NS \quad T=AP/2$$

(b) Using those formulas, write a program **EX143B** that accepts **N** and **R** and displays **P** and **T**.

(c) Add a step to your program to round your values of **P** and **T** to two decimal places.

(d) Modify the formulas in part (a) appropriately to develop a program **EX143D** that accepts **N** and **S** and calculates **P** and **T**.

(14.4) One way to find out what a program is doing is to compare input with output for a number of values. Here is a program that produces a rather strange sequence of values.

PROGRAM:EX144

```
Prompt X
1→A
1→B
0→Y
While X≥A
   Y+1→Y
   B+2→B
   A+B→A
End
Disp Y
```

(a) Enter and run that program for **X** values from **1** to **10**.

(b) Describe in your own words what you think the program is doing.

(c) Check your answer to (b) with several additional values.

(d) This program is executing a combination of two of the following functions, **2X**, **X/2**, **X2**, √**X**, **abs**, **int**, **max** and **min**. Add a line at the end of your program **EX144** that displays a combination you think might work. For example, your extra line might read **Disp min(2X)**. Then run the program: if your choice is correct, the program should display two equal values. Adjust your choice until you achieve this.

(14.5) Credit card companies require minimum payments that usually amount to about 2% per month or the entire amount if the debt is $15 or less. They also charge a current average of 14.7% per year or 1.225% per month. (If you fail to make a required payment, you become a credit risk and the annual rate rises to about 24%.) As of this writing, the average credit card debt carried by a recent college graduate is $4100.

(a) If you owe \$4100 and pay \$100 per month at 1.225% per month interest charge, how many months would it take you to pay off that loan and how much would you have had to pay?

(b) Answer the same questions if you pay only the minimum amount due each month until you owe \$80 or less, when you pay that amount. (Note that the first month that payment would be \$82, almost as much as the \$100 paid in part (a).)

(14.6) An interesting algorithm was discovered by the German mathematician Lothan Collatz in 1937. Collatz introduced a simple algorithm that produces a sequence of positive integers:

1. Begin with any positive integer, X.
2. If X is even, the next number is $X/2$; if X is odd, the next number is $3X+1$.
3. Repeat step 2 on the latest number until you reach 1.

Here are two examples of sequences produced starting with 3 and 9:

$3 \rightarrow 10 \rightarrow 5 \rightarrow 16 \rightarrow 8 \rightarrow 4 \rightarrow 2 \rightarrow 1$

$9 \rightarrow 28 \rightarrow 14 \rightarrow 7 \rightarrow 22 \rightarrow 11 \rightarrow 34 \rightarrow 17 \rightarrow 52 \rightarrow 26 \rightarrow 13 \rightarrow 40 \rightarrow 20 \rightarrow 10 \rightarrow 5 \rightarrow 16 \rightarrow 8 \rightarrow 4 \rightarrow 2 \rightarrow 1$

Collatz conjectured that no matter what starting positive integer is chosen, the sequence will always end at 1. As of 2010, this conjecture was not proved and a £2000 prize has been offered for either a counterexample (a sequence that does not reach 1) or a proof that all sequences reach 1.[2] Clearly that is not a reasonable exercise for you to address. You can, however, develop the program **COLLATZ** that will generate the sequence. The following steps will help you to do that. I suggest that you write out the program lines on paper before entering them in your calculator.

(a) Write a program line that carries out step (1) of the algorithm.

(b) Develop an **If...Then...Else...End** series of program lines that carries out algorithm steps (2) and (3). A test that determines if a number is even is **X/2=int(X/2)**.

(c) You are going to want to repeat those steps until you reach 1. You can do that by including the structure you developed in a **While...End** loop. Here the test governing the loop will be **X≠1**.

(d) You have completed the process, but you have not displayed any results. You will want to do this immediately after your **If...Then...Else...End** loop. Add a **Disp** line here as well as a following **Pause** line so that your calculator will stop for you to see each value.

(e) Test your program with the values **3** and **9** that were worked out as examples.

(14.7) Construct the program **COLLATZ2** that will count the number of values in the sequence for a given number. These steps will help.

(a) Once you have named the new program, use **RCL PRGM EXEC** to copy the steps in **COLLATZ** into **COLLATZ2**.

(b) Using **C** for your counter, you will need to insert a program line **1→C** after you prompt for your initial **X**.

(c) You won't want to display each value of the sequence so you can delete the two lines that recorded them. Replace those lines with a line that increments **C**.

(d) At the end of the program enter a line that displays the final value of **C**.

(e) Test your program on the following numbers: **3**, **9**, **27** and **837799**. What number less than 100 leads to the highest number of steps?

(14.8) You can reduce the number of steps in the sequence by making a small adjustment to your programs **COLLATZ** and **COLLATZ2**. Replace **3X+1→X** with **(3X+1)/2→X**.

(a) Test several values to assure yourself that, when X is odd, 3X+1 is always even.

(b) Argue that (3X+1)/2 is always an integer.

(c) Make these program changes and compare your results for the initial numbers that you used in (14.7e).

(14.9) It may seem clear to you from your experimenting with this Collatz algorithm that all numbers will finally reach 1. In fact, as I write this, that has been established for all positive numbers up to 5.64×10^{18}. But you need only consider negative numbers to see other possibilities. For example, **−5** leads to the sequence **−5**, **−14**, **−7**, **−20**, **−10**, **−5** at which point it continues in that loop. Find other negative numbers for which the sequence does not reach **−1**. (Use the program **COLLATZ** for this experimentation.)

- - - - -

2 The popularity of this problem is indicated by the many other names by which it is known. Some of them are the 3n+1 problem, Hasse's algorithm, Kakutani's problem, Thwait's conjecture, the Syracuse algorithm and Ulam's problem.

- - - - -

(14.10) The cells in a prison are numbered 1-100. The jailer controls the cell locks electronically at a control board. He receives a series of strange instructions which he dutifully carries out. First, he is told to unlock all the cell doors. Next he is told to relock every second door, that is, the doors numbered 2, 4, 6,... Then he is told to change the locks of every third door (those numbered 3,6,9,...): if the door is locked, unlock it; if unlocked, lock it. These lock-switching instructions continue for every 4th door, every 5th door, all the way up to switching the 100th door. At the end of this process which doors will be unlocked? Develop a program **EX1410** to answer this question.

(a) This is a perfect exercise on which to apply lists. You can set up the original status of the cells, with all the cells locked (assuming **1** for locked, **0** for unlocked), with these program lines:

```
100→dim(L1)
Fill(1,L1)
```

(b) Now you'll need two **For...End** loops, an outer one to control the number, **N**, of passes and an inner one that controls the cells to be switched. The loop control **For(C,N,100,N)** will work for this. When **N** = **5**, what values of **C** will this loop process?

(c) Within these two nested loops you will want to make the switches. Do that with an **If...Then...Else...End** structure.

(d) If you have programmed (a) through (c) correctly, you should have a list indicating which doors are open and which closed. That list should begin like this: **{0 1 1 0**

`1 1 1 1 0 1 1 1` .... Now you want to construct another list, L_2, that indicates which cells are open. You'll need to initiate that second list and use another **For...End** loop to assign values. You'll also need a variable, **K** to keep track of which L_2 element to be filled. You can start this process with these three lines:

```
1→K
For(C,1,100)
  If L1(C)=0
    . . .
```

(e) One important use of a programmed solution to a problem is as a tool for rethinking that problem. In this case the solution represents a particular kind of number. Something that may not have occurred to you in working out this exercise is the fact that the cells that are unlocked are those with an odd number of factors. Do some experimenting to determine which numbers have an odd number of factors?

Chapter 15.
Exercise Answers

15.1 Chapter 1

No exercises.

15.2 Chapter 2 Answers

(2.1) (a) 1.25   (b) .8   (c) 7396   (d) 625   (e) .0009765625

  (f) 15.70796327   (g) .6020599913   (h) 148.4131591

(i) 4

(2.2) A = 4, B = 4, C = 5, and D = 4

(2.3) (a) -   (b) (-)   (c) (-), -

(2.4) (a) 1.782451739   (b) 1.732050808   (c) 1.029098992

  (d) .029... < .03 = 3%

(2.5) The command **a EE b** displays scientific notation, aEb for

  $a \times 10^b$

  (a) 10   (b) 873000   (c) .0023

15.3 Chapter 3 Answers

(3.1) (a) 98.6°   (b) -459.67°   (c) 536°

  (d) 10832°   (e) 224.6°   (f) -243.4°

(3.2) -40°

(3.3) (a) `PROGRAM:FTOC`
  `Prompt F`

```
        5/9*(F-32)→C
        Disp C
```

(b) $98.6 \rightarrow 37$, $-459.67 \rightarrow -273.15$

(3.5) `PROGRAM:NAME`
```
      Disp "your name"
      PROGRAM:TRIGSUM
      Prompt X
      sin(X)+cos)X→Y
      Disp Y
```

(a) 1.311235982   (b) -1.328926049   (c) 1

15.4 Chapter 4 Answers

(4.1) `PROGRAM:NAME`
```
      Disp "your name"
      Disp "OWNER"
```

(4.2) `PROGRAM:FTOC`
```
       Input"FTEMP ",F
       5/9(F-32)→C
       ClrHome
       Output(2,1,"FTEMP")
       Output(3,1,F)
       Output(3,5,"DEGREES")
       Output(5,1,"CTEMP")
       Output(6,1,C)
       Output(6,5,"DEGREES")
```

15.5 Chapter 5 Answers

(5.1) (b) The line **Disp "CONVERT TEMPS"** is outside the **Lbl A...Goto A** loop; therefore it does not appear after the first time through the program.

(5.2) The program will follow your instructions and print your name and **OWNER** over and over, but it will do this without stopping and you will hardly see them as they pass. If you add **Pause** after the display lines, the program would stop at each pass and allow you to read your two lines.

15.6 Chapter 6 Answers

(6.1) **X=Y** is a test and **X→Y** stores the value of **X** in **Y**.

(6.2) (a) 0   (b) 1   (c) 1   (d) 0   (e) 0   (f) 1 (if in Degree mode)

(6.3) No

(6.4) (a) **iPart(23.35)** is **23** and **fPart(23.35)** is **.35**. **iPart(** is the integer part and **fPart(** is the fraction or decimal part of a given number.

(b) (1) 3   (2) 5   (3) 5   (4) 14

(c) The positive errors have accumulated.

(d) It does give the same value. Adding .5 to a number will make it more than the next integer only when the original number is half way or more already. Then **iPart(** strips off the decimal.

(6.5) (a) **91** would make the first If true. You would display **HONORS** and skip the rest of the program.

(b) **56** would take you to the first **Else** and then within that to the second **Else**. You would dutifully display **FAIL**.

15.7 Chapter 7 Answers

(7.1) (a) 3.76   (b) 3.76   (c) 3   (d) -4   (e) -3

   (f) 2   (g) .65   (h) -.65   (i) .142

 (7.2) The answers are all the same. The two commands are equivalent.

   Results: (a) .7   (b) -.46   (c) -.1   (d) .8   (e) .3535

(7.3)   **PROGRAM:TEST**
```
Prompt A
Disp round(A,0)
Disp int(A+.5)
```

 (a) 24   (b) -16   (c) -2   (d) 18   (e) 0

(7.4) Running the program for the given assignments of **N** and **D** gives:

 (a) 8   (b) 7   (c) 5   (d) 4

The program is identifying the digit in the number **N** that is **D** places from the decimal point, positive to the left, negative to the right.

(7.5) **A** is an integer.

(7.6) (a) integers between 0 and 9 inclusive   (b) 0 and 1

   (c) integers between 0 and 99 inclusive

(7.7) (a) **randInt(3,7)**

   (b) **2randInt(1,6)**

   (c) **2randInt(1,5)-1**

or **2randInt(0,4)+1**

(d) **3randInt(11,33)**

(7.8) **randInt(1,10)**

15.8 Chapter 8 Answers

(8.1) (a) 2, 3, 4, 5, 6   (b) 231, 232, 233, 234

(c) -3, -2, -1, 0   (d) 35

(8.2) (a) 5, 7, 9, 11   (b) 13, 20, 27   (c) 9, 8, 7, 6, 5   (d) 3, 1, -1, -3

(8.3) (a) **For(X,0,360,30)**

(b) You will need a Pause in order to read each value.

```
PROGRAM:TRIGSUM
For(X,0,360,30)
   sin(X)+cos(X)→Y
   Disp X,Y
   Pause
End
```

(8.4)   PROGRAM:SUMN
```
0→S
    For(N,1,100)
   S+N→S
End
Disp S
```

(8.5) (a) 3025   (b) 1.428968254   (c) 61.66597781

        PROGRAM:CUBESUM
```
0→S
For(I,1,10)
```

```
    S+I^3→S
End
Disp S
```

```
PROGRAM:RECIPSUM
0→S
For(I,3,10)
   S+1/N→S
End
Disp S
```

```
PROGRAM:SQRTSUM
0→S
For(I,3,10)
S+√N→S
End
Disp S
```

(8.6) (a)
```
PROGRAM:NAME
Lbl 1
Disp"XY"
For(N,1,2000)
End
Goto 1
```

(b) various values: I got 29.5 sec which gives 678 times per second

(c) Good approximation.

(8.7) (a)   (b) **A** is **17**   (c) **B** is **6**

line	A	B
2	5	

3	5	3
4	8	3
3	8	4
4	12	4
3	12	5
4	17	5
6	17	6

15.9 Chapter 9 Answers

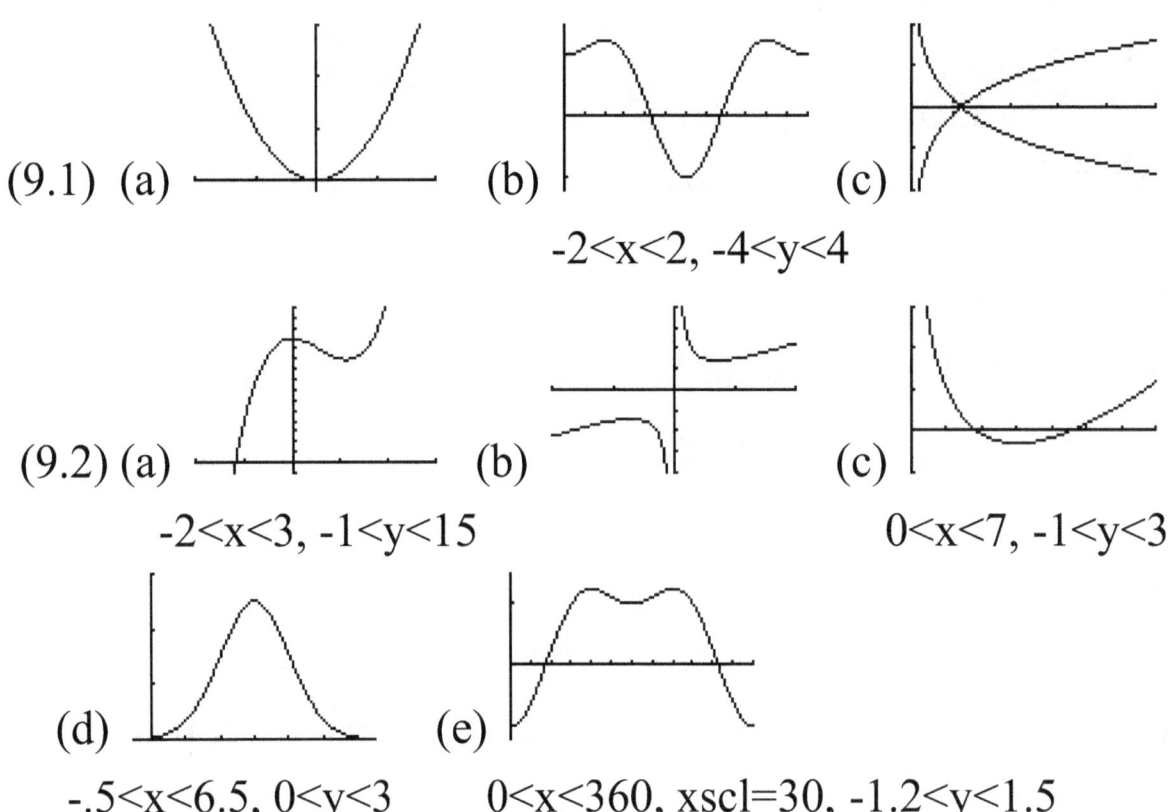

(9.1) (a)   (b)   (c)

-2<x<2, -4<y<4

(9.2) (a)   (b)   (c)

-2<x<3, -1<y<15   0<x<7, -1<y<3

(d)   (e)

-.5<x<6.5, 0<y<3   0<x<360, xscl=30, -1.2<y<1.5

(9.3) (a) an ellipse   (b) Apparently. X-range 6; Y-range 4

(c) Here are the answers as they relate to the program line that produces the X and Y values.

line	X	B

5	10	10
6	10	50
7	10	10
8	50	10
5	11	10
6	11	50
7	10	11
8	50	11
5	12	10
6	12	50

The pixel (10,10) appears twice, as will (50,50) later.

(9.4) (a) across, then down

(b) (62,90),(62,91),(62,92),(62,93),(62,94)

(c) down, then across

(9.5) (a) This 15-pixel **N** is easily confused with the letters in (b). If the connecting pixel were simply placed at the top, it would look like an upside-down U. Note that the resulting choice looks a bit like a lower case n.

(b) These letters differ only in the pixel between the side bars.

(c) Comparing the15-pixel and the 35-pixel **N**s

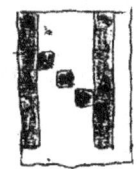

(d) The Home screen will allow 16 characters (letters or numbers) across and 8 down. The Pixel screen will allow 23 letters or numbers across and 10 down with a few pixels to spare in each direction.

15.10 Chapter 10 Answers

(10.1) (a)
```
PROGRAM:EXREP101
  8→N
  Repeat N>10
    Disp N
    N+1→N
  End
```

(b)
```
PROGRAM:EXFOR101
For(N,8,10)
Disp N
End
```

(10.2) (a) 0 and any negative number

(b) The exercise is about positive numbers, so 0 is lower than any possible number to be entered Thus the first entry will establish a high value and the program will continue from there.

(10.3) (a) If **D** is not defined, your calculator will give you an error message. If it has a value from previous calculations, that value will be used. There will be no problem unless that value is zero or negative, in which case the loop will not be evaluated.

(b) `Repeat D≤0`

(10.4) (a) For **N** = **3**, the program would display just **3**. For **4**, it would display **4** and **5**.

(b) Because the **Repeat...End** loop is checked at the end and is always evaluated at least once, you would display **5** and then, because **N** would become **6**, the loop terminates.

15.11 Chapter 11 Answers

(11.1) 8

(11.2) 9

(11.3) Here are full and abbreviated forms for this answer:

command	A	B	C	D
initial values	1	2	3	
A+C→D	1	2	2	4
$C^2 + D^2 \rightarrow A$	25	2	3	4
$\sqrt{A} \rightarrow B$	25	5	3	4
(A + B + C) / C → D	25	5	3	11
$\text{int}\left(\sqrt{B}\right) \rightarrow C$	25	5	2	11
If A < 7 : Then : 3 → D	25	5	2	11

command	A	B	C	D
initial values	1	2	3	
A+C→D				4
$C^2 + D^2 \rightarrow A$	25			
$\sqrt{A} \rightarrow B$		5		
(A + B + C) / C → D				11
$\text{int}\left(\sqrt{B}\right) \rightarrow C$			2	
If A < 7 : Then : 3 → D				

(11.4) (a) For the values **A** = 3, **B** = 5

line	A	B	T
1	3	5	
8	3	5	

(b) For the values **A** = 5, **B** = 3

line	A	B	T
1	5	3	
4	5	3	5
5	3	3	5
6	3	5	5
8	3	5	5

(11.5) For the values **A** = 5, **B** = 3

line	A	B
1	5	3
4	3	3
5	3	3
7	3	3

(11.6) Simply change the test in line 2 from **If A>B** to **If**

A<B.

(11.7) (a) For **A = 1, B = 3, C = 2**

line	A	B	C	T
1	1	3	2	
10	1	3	2	3
11	1	2	2	3
12	1	2	3	3
20	1	2	3	

(b) For A=3, B=1, C=2

line	A	B	C	T
1	3	1	2	
4	3	1	2	3
5	1	1	2	3
6	1	3	2	3
10	1	2	2	3
11	1	2	3	3
12	1	2	3	
20	1	2	3	

(c) For A=3, B=2, C=1

line	A	B	C	T
1	3	2	1	

103

4	3	2	1	3
5	2	2	1	3
6	2	3	1	3
10	2	3	1	3
11	2	1	1	3
12	2	1	3	3
16	2	1	3	2
17	1	1	3	2
18	1	2	3	2
20	1	2	3	

(11.8) Lines 15-20 were necessary to process case (c) in (11.7).

15.12 Chapter 12 Answers

(12.1) PROGRAM:EX121
```
{2,2}→dim([B])
7→[B](1,1)
9→[B](1,2)
1→[B](2,1)
5→[B](2,2)
```

(12.2) PROGRAM:EX122
```
{1,10}→dim([C])
For(N,1,10)
  N→[C](1,N)
End
```

(12.3) PROGRAM:EX123

```
{3,10}→dim([D])
For(N,1,10)
N→[D](1,N)
7N→[D](2,N)
5(10-N)→[D](3,N)
End
```

(12.4) (a) **PROGRAM:EX124A**
```
{1,10}→dim([E])
For(N,1,10)
  √(N)→[E](1,N)
End
```

    (b) **PROGRAM:EX124B**

```
Prompt X
```

```
Disp[E](1,X)
```

15.13 Chapter 13 Answers

(13.1) (a) Unless you have defined a value of **Z**, you will get `{2,0}`.

(b) You have stored the value **2** in **A** so that is the first entry. Undefined entries appear as **0**s.

(13.2) Lists:

  (a) `{3 6 9 12}` (b) `{5 5 5 5}`

  (c) `{1 4 9 16}` (d) `{4 6 6 4}`

  (e) `{4 3 2 1 1 2 3 4}`

  (f) `{4 3 2 1}`

(13.3) `{1 4 9 16 25 36 49 64 81 100}`

(13.4) (a) Use **"*your name*"→**`Str2`

```
PROGRAM:EX134A
Output(4,4,"HELLO")
Output(4,10,Str2)
```

    (b) **PROGRAM:EX134B**
```
prgmPXLSETUP
Text(25,35,"HELLO ",Str2)
```

    (c) Use **"HELLO "→**`Str3`

```
PROGRAM:EX134C
Output(4,4,Str3+Str2)
```

    (d) **PROGRAM:EX134D**
```
prgmPXLSETUP
Text(25,35,Str3+Str2)
```

15.14 Chapter 14 Answers

(14.1)

line	X	Y	P	test
1	25	21		
2			0	
3				true
4				false
8		20		
9			25	
3				true
4				true

106

line	X	Y	P	test
5		10		
6	50			
4				true
5		5		
6	100			

line	X	Y	P	test
4				false
8		4		
9			125	
3				true
4				true
5		2		
6	200			
4				true
5		1		
6	400			
4				false
8		0		
9			525	
3				false
11			525	

(14.2) It will take 47 years from the current year.

```
PROGRAM:EX142
227→U
70→M
2012→Y
While U>M
   U+.007U→U
   M+.033M→M
   Y+1→Y
End
Disp Y
```

(14.3) (b) `PROGRAM:EX143B`
```
Prompt N,R
Degree
180/N→X
Rsin(X)→S
Rcos(X)→A
NS→P
AP/2→T
Disp P,T
```

   (c) Add the program lines **round(P,2)** and **round(T,2)** before the final line.

   (d) Using $A = S/(2\tan(X))$ and $T = AP/2$:

```
PROGRAM:EX143D
Prompt N,S
Degree
180/N→X
S/(2tan(X))→A
NS→P
AP/2→T
```

```
Disp P,T
```

(14.4) (a) The values produced for 1-10 are: **1 1 1 2 2 2 2 2 3 3**.

　　　(b) The sequence begins with **1** and increases by **1** at 3 and 8.

　　　(d) The correct command is **int($\sqrt{}$(X))**.

(14.5) (a) You would pay \$5730.10 in 58 months.

```
PROGRAM:EX145A
4100→A
0→M
0→T
While A>0
T+100→T
A+.01225A-.02A→A
M+1→M
End
Disp M
Disp round(T+A,2)}
```

　　　(b) You would pay \$10,136.84 in 491 months. (The reason for the long payment period is the small amount paid off each month during the later months. If you waited to make the final payment of the required \$15, you would have paid \$10,233.89 in 699 months.

```
PROGRAM:EX145B
4100→A
0→M
0→T
While A>80
```

```
T+.02A→T
A+.01225A-.02A→A
M+1→M
End
Disp M
Disp round(T+A,2)}
```

(14.6) (a) **Prompt X**

(b) **If X/2=int(X/2)**
```
Then
   X/2→X
Else
   3X+1→X
End
```

(d)   **PROGRAM:COLLATZ**
```
Prompt X
While X≠1
If X/2=int(X/2)
Then
   X/2→X
Else
   3X+1→X
End
Disp X
Pause
End
```

(14.7) (a)   **PROGRAM:COLLATZ2**
```
Prompt X
X→Y
1→C
```

```
While X≠1
If X/2=int(X/2)
Then
   X/2→X
Else
   3X+1→X
End
C+1→C
End
Disp Y,C
```

   (e) Steps for 3 : 8; 9 : 20; 27 : 112; 837788 : 525. The most steps for numbers less than 100: `97: 119`.

(14.8) (b) 3 times an odd number is odd. Add one and the number is even. Any even number is divisible by 2.

   (c) This simple change saves about 1/3 of the steps.

N	old	new
3	8	6
9	20	14
27	112	71
837799	525	330
97	119	76

(14.9) Of course, any number that is -5×2^n will reduce to -5 and enter the same loop. Other numbers include -7, -9, -13 and many more. The number `-17` leads to a much longer loop.

 (14.10) The final program can be:

```
PROGRAM:EX1410
100→dim(L₁)
Fill(1,L₁)
For(N,1,100)
   For(C,N,100,N)
     If  L₁(C)=0
     Then
         1→L₁(C)
     Else
         0→L₁(C)
      End
   End
End
20→dim(L2)
1→K
For(C,1,100)
   If  L₁(C)=1
   Then
      C→L₂(K)
      K+1→K
   End
End
```

Once you have run this program you can key L_2 to see the numbers of the open lockers.

 (d) All numbers except squares have an even number of factors and the locks are changed with each factor.

www.ingramcontent.com/pod-product-compliance
Lightning Source LLC
Chambersburg PA
CBHW081135170526

45165CB00008B/2679